Steel

Steel Design covers steel design fundamentals for architects and engineers, such as tension elements, flexural elements, shear and torsion, compression elements, connections, and lateral design. As part of the *Architect's Guidebooks to Structures* series it provides a comprehensive overview using both imperial and metric units of measurement. Each chapter includes design steps, rules of thumb, and design examples. This book is meant for both professionals and for students taking structures courses or comprehensive studies. As a compact summary of key ideas, it is ideal for anyone needing a quick guide to steel design. More than 150 black and white images are included.

Paul W. McMullin is an educator, structural engineer, and photographer. He holds degrees in Mechanical and Civil engineering and is a licensed engineer in numerous states. He is a founding partner of Ingenium Design, providing innovative solutions to industrial and manufacturing facilities. Currently an adjunct professor at the University of Utah in Salt Lake City, USA, he has taught for a decade and loves bringing project-based learning to the classroom.

Jonathan S. Price is a structural engineer and adjunct professor at Philadelphia University in Pennsylvania, USA, where he was honored with the Distinguished Adjunct Faculty Award in 2006. He holds a Bachelor of Architectural Engineering degree from the University of Colorado, USA, a Master of Science degree in civil engineering from Drexel University in Philadelphia, USA, and is registered in 12 states.

Richard T. Seelos, SE, LEED AP, is a structural engineer who has designed large commercial office buildings, hospitals, educational buildings, and heavy industrial structures. He gained a different perspective while working for a contractor as the lead structural quality control manager on a 1.2-billion-dollar project. He also taught the Introduction to Steel and Concrete course for several years as an adjunct professor at the University of Utah. He currently works as a project manager for Bennett & Pless in Atlanta, GA.

Architect's Guidebooks to Structures

The *Architect's Guidebooks to Structures* series addresses key concepts in structures to help you understand and incorporate structural elements into your work. The series covers a wide range of principles, beginning with a detailed overview of structural systems, material selection, and processes in *Introduction to Structures*; following with topics such as *Concrete Design*, *Special Structures Topics*, *Timber Design*, and *Steel Design* to equip you with the basics to design key elements with these materials and present you with information on geotechnical considerations, retrofit, blast, cladding design, vibration, and sustainability.

Designed as quick reference materials, the *Architect's Guidebooks to Structures* titles will provide architecture students and professionals with the key knowledge necessary to understand and design structures. Each book includes imperial and metric units, rules of thumb, clear design examples, worked problems, discussions on the practical aspects of designs, and preliminary member selection tables; all in a handy, portable size.

Read more in the series blog: http://architectsguidestructures.wordpress.com/

Concrete Design
Paul W. McMullin, Jonathan S. Price, and Esra Hasanbas Persellin

Introduction to Structures
Paul W. McMullin and Jonathan S. Price

Timber Design
Paul W. McMullin and Jonathan S. Price

Special Structural Topics
Paul W. McMullin, Jonathan S. Price, and Sarah Simchuk

Steel Design
Paul W. McMullin, Jonathan S. Price, and Richard T. Seelos

Masonry Design
Paul W. McMullin and Jonathan S. Price

Steel Design

Edited by Paul W. McMullin,
Jonathan S. Price and
Richard T. Seelos

NEW YORK AND LONDON

First published 2018
by Routledge
711 Third Avenue, New York, NY 10017

and by Routledge
2 Park Square, Milton Park, Abingdon, Oxon OX14 4RN

Routledge is an imprint of the Taylor & Francis Group, an informa business

Library of Congress Cataloguing-in-Publication Data
A catalog record for this book has been requested

ISBN: 978-1-138-83104-9 (hbk)
ISBN: 978-1-138-83106-3 (pbk)
ISBN: 978-1-315-73680-8 (ebk)

Acquisition Editor: Wendy Fuller
Editorial Assistant: Kalliope Dalto
Production Editor: Alanna Donaldson

Typeset in Calvert
by Florence Production Ltd, Stoodleigh, Devon, UK

Printed and bound in the United States of America by Sheridan

For our kids
Ruth, Peter, and Katherine (Beast)
Leslie and Deirdre
Alan, Duncan, Cooper, Afton, and Parker

Contents

Acknowledgments

Like previous editions, this book wouldn't be what it is without the diligent contributions of many people. We thank Sarah Simchuk for her wonderful figures and diligent efforts; Phil Miller for reviewing each chapter; Bill Komlos for his helpful insight on welding; and Kevin Churilla for his work on the example figures.

We thank Wendy Fuller, our commissioning editor, Alanna Donaldson, our production editor, Quentin Scott, our copy editor, and Natasha Gibbs at Florence Production. Each of you have been wonderful to work with and encouraging and helpful along the journey. We thank everyone at Routledge who produced and marketed the book.

A special thanks to our families, and those who rely on us, for being patient when we weren't around.

We are unable to fully express our gratitude to each person involved in preparing this book. It is orders of magnitude better than it would have otherwise been thanks to their contributions.

Contributors

EDITORS

Paul W. McMullin, SE, PhD, is an educator, structural engineer, and photographer. He holds degrees in Mechanical and Civil engineering, and is a licensed engineer in numerous states. He is a founding partner of Ingenium Design, providing innovative solutions to industrial facilities. Currently an adjunct professor at the University of Utah in Salt Lake City, USA, he has taught for a decade and loves bringing project-based learning to the classroom.

Jonathan S. Price, PE, LEED AP, His journey over the last 40 years has been in building construction or design and education. Armed with a Bachelor of Architectural Engineering degree from the University of Colorado in 1977 and a Master of Science degree in Civil Engineering from Drexel University in Philadelphia in 1992, Mr. Price has worked in various capacities in a variety of design firms. He has taught structural design at Philadelphia University since 1999 and was honored with the Distinguished Adjunct Faculty Award in 2006.

Richard T. Seelos, SE, LEED AP, is a structural engineer who has designed large commercial office buildings, hospitals, educational buildings, and heavy industrial structures. He gained a different perspective while working for a contractor as the lead structural quality control manager on a 1.2-billion-dollar project. He also taught the Introduction to Steel and Concrete course for several years as an adjunct professor at the University of Utah. He currently works as a project manager for Bennett & Pless in Atlanta, GA.

CONTRIBUTORS

Kevin S. Churilla is an industrial designer has drafted and 3-D modeled for two decades. His years of framing experience make his drawings simple and clear. He is a founding partner of Ingenium Design. He and his family enjoy soccer, camping, fishing, and going to hockey games together.

Sarah Simchuk is a Project Architect and fine artist working towards architectural licensure in large-scale retail design. She holds bachelor's and master's degrees in architecture from the University of Utah. She is in the early stages of her architectural career, with an inclination towards design and details in project management. She comes from a fine art background with over 15 years' experience in hand drawing and rendering, and lends a 3-D approach to the understanding of structures.

Introduction

The advent of structural steel has profoundly changed building and bridge design across the globe. Rising from cast iron technology, steel makes possible skyscrapers and long-spanning bridges, informing our cityscapes. It is found in homes, warehouses, civic buildings, and industrial facilities, and is one of the basic building blocks of today's infrastructure.

This guide is designed to give the student and budding architect a foundation for successfully understanding and incorporating steel in their designs. It builds on *Introduction to Structures* in this series, presenting the essence of what structural engineers use most for steel design.

If you are looking for the latest steel trends, or to plumb the depths of technology, you're in the wrong place. If you want a book devoid of equations and legitimate engineering principles, return this book immediately and invest your money elsewhere. However, if you want a book that holds architects and engineers as intellectual equals, opening the door of steel design, you are very much in the right place.

Yes, this book has equations. They are the language of engineering. They provide a picture of how structure changes when a variable is modified. To disregard equations is like dancing with our feet tied together.

This book is full of in-depth design examples, written the way practicing engineers design. These can be built upon by reworking examples in class with different variables. Better yet, assign small groups of students to rework the example, each with new variables. Afterward, have them present their results and discuss the trends and differences.

For learning assessment, consider assigning a design project. Students can use a past studio project, or a building that interests them. The project can start with determining structural loads, continue with designing key members, and end with consideration of connection and seismic design. They can submit design calculations and sketches summarizing their work and present their designs to the class. This approach requires a basic level of performance, while allowing students to dig deeper into areas of interest. Most importantly it places calculations in context; providing an opportunity to wrestle with the iterative nature of design, and experience the discomfort of learning a new language.

Our great desire is to bridge the gap between structural engineering and architecture. A gap that historically didn't exist, and is unnecessarily wide today. This book is authored by practicing engineers, who understand the technical nuances and the big picture of how a steel project goes together. We hope it opens the door for you.

User note: This book makes extensive use of the AISC section properties. A handful of the available sections are in Appendix 1. However, there may be times that having access to the full set of sections and properties will be helpful. AISC currently makes these available on their website at www.aisc.org/publications/steel-construction-manual-resources/#28293

LeBow Business Building

Chapter 1

Jonathan S. Price

Figure 1.1 LeBow Building

Source: Keast & Hood

The construction of any significant work requires hard work both intellectually and physically. Honesty, adaptability, and attention to detail help also. The LeBow School of Business is a case in point. From its outset, the project was a challenge to the entire team. The only predictable outcome was that a new building would be constructed. Its appearance, size, shape, and final cost would be determined by those who would mold the clay that started as an idea, and finished with an opening that belied the struggle we all went through to see it completed.

1.1 INTRODUCTION

Bennett LeBow is a Drexel University Graduate and a Philadelphia native. His generosity provided the major funding necessary to construct the LeBow College of Business. When the project was announced, several well-known architects competed for the commission. The winning team was Robert A. M Stern with Voith-MacTavish Architects.

The proposed site was occupied by Matheson Hall, a depressing, four-story, 60s-style structure facing Market Street. Before Matheson was constructed, row houses and small buildings lined both Woodland Avenue and Market Streets.

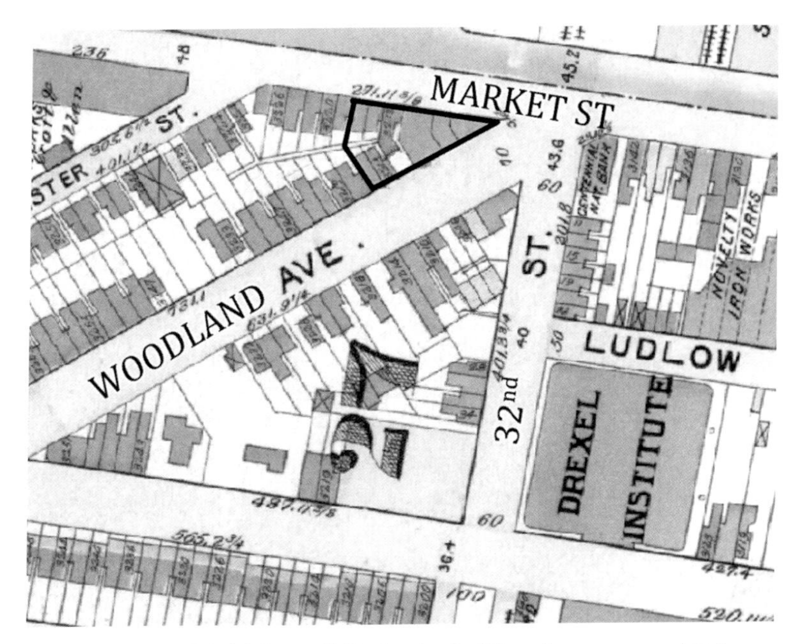

Figure 1.2 West Philadelphia c. 1895, showing building site names annotated

Source: Philageohistory.org

Incorporating Matheson Hall was an early goal but its restrictive headroom, shallow basement, narrow **bays**, and small foundations limited the new design. Reinforcing the structure and living with the existing column grid would negate any savings and compromise the program. Matheson had to go.

About 30 ft (9.1 m) below Market Street is Philadelphia's main east–west commuter line, the Market–Frankford EL (elevated). This subterranean structure presented challenges but helped in other ways as will be discussed. Other non-yielding site features included public utilities within the Market and Woodland Avenue rights of way. Taken together, these restrictions defined a tight triangular property, shown in Figure 1.2.

Reflecting the site, a triangular plan evolved with symmetry about the long bisector. The two long sides or "bars" define a central, 7-story-high atrium. The shortest bar is the tallest at 12 stories and houses faculty offices and classrooms. The other two bars are about half as high at

7 stories each and support classrooms, large meeting rooms, and a very large, lower level auditorium.

1.2 DESIGN DEVELOPMENT

With an aggressive schedule and many unknowns, a steel **frame** was the logical choice for its inherent **strength** and design flexibility. Steel costs were disproportionately high and the structural engineers looked for savings. After the initial modeling was complete, the steel weighed-in at approximately 18 psf, including **columns**. This was too heavy so we explored ways to reduce the weight. Some of this high weight was attributable to wide bays, initial conservatism, and an excessive number of heavy column transfers required by floor plan alignment (or misalignment) between levels.

Value Engineering (VE), often bemoaned by design teams, is a CM-driven process to identify potential savings used to meet a construction budget. On LeBow, VE offered difficult choices. One of the least palatable was to shell some of the tower office levels pending added funding for fit-out (shell means build the space without finishes).

Motivated by the cost pressure, the architects reworked the plans to better align walls (and columns) between floors. Some column transfers were inevitable, such as those above the large subterranean auditorium. We worked on reducing our initial conservatism on the structure.

Other challenges that surfaced were below grade. Based on groundwater observation wells, the geotechnical engineer recommended we design for 8 feet (2.44 m) of hydrostatic pressure beneath the lower level. This impacted the site development costs and drove the design decision to use a mat foundation system acting as a very large boat. A blessing in disguise, the Market Street subway with its own system of drains, helped lower the groundwater by acting as a large French drain during construction.

Our solution for spanning over the lower level auditorium was to design story-high **trusses** extending from the 2nd to the 3rd floors to be concealed behind the atrium walls (Figure 1.3). Temporary columns **supported** the first, second, and third floor framing until the trusses were built and stabilized. The columns between levels one and two, below the trusses, were initially in **compression** but became hangers after the trusses were put into service.

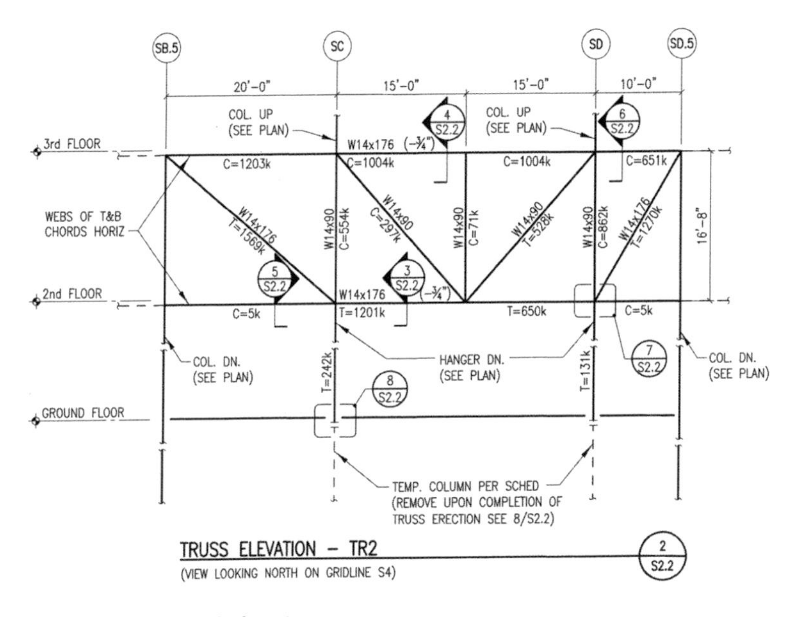

Figure 1.3 Truss TR-2 Elevation

Source: Keast & Hood

The trusses above the auditorium were modeled in STAAD® as simply supported plane trusses with one top **chord** support pinned and the other on a **roller**. The bottom chords were modeled with roller or slide supports.

Cantilevers, another device used to convey weightlessness, spawned the question "how thin can you make it?" As the engineers for many architecturally driven projects, we try hard to provide an economical solution but are mindful of the architects' wishes. This came into play on the eastern façade where we absorbed the structure into the slab, saving 5 in (127 mm) of **depth**, shown in Figure 1.5. This was at no small cost from an engineering and construction standpoint, but was important architecturally.

At the northwest corner adjacent to Market Street, a cantilevered corner was structured with a large diagonal rod that would remain visible behind the aluminum **curtain wall** and became a design feature, shown in Figure 1.6. We cantilevered orthogonal W24 × 104 from the nearest point on line N1 to provide needed redundancy and **stability** during

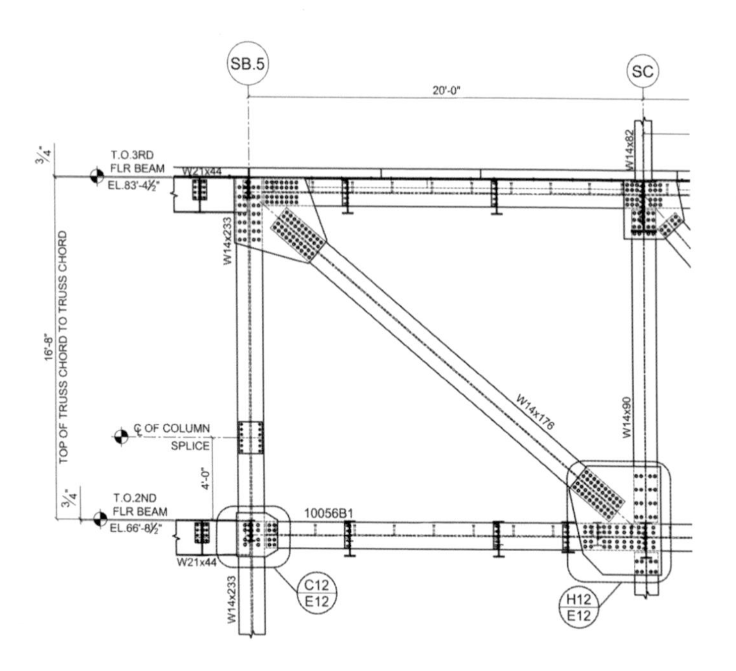

Figure 1.4 Truss TR-2 Column Connections

Source: Keast & Hood

erection until the 2 in (51 mm)-diameter diagonal rod was installed and tensioned.

Resisting lateral **loads** presented a larger challenge. Surrounding the elevator and stair **core** and rising some 250 ft (46 m) above the foundations, the tower had its own diagonally **braced** core. Because the building was to be relatively long in the east–west direction, the broader 7-story section off to the east demanded a lateral system of its own. For that, we designed concrete **shear walls** around the easternmost stair rising 7 stories.

The two bracing systems responded very differently. In our model; the steel core bracing acted as a tall-cantilevered truss and the shear walls as squat, rigid **elements** of the structure. The shear walls attracted tremendous **forces** from the tower bracing system. Rather than separating the two systems with an **expansion joint**, as would be required in a high

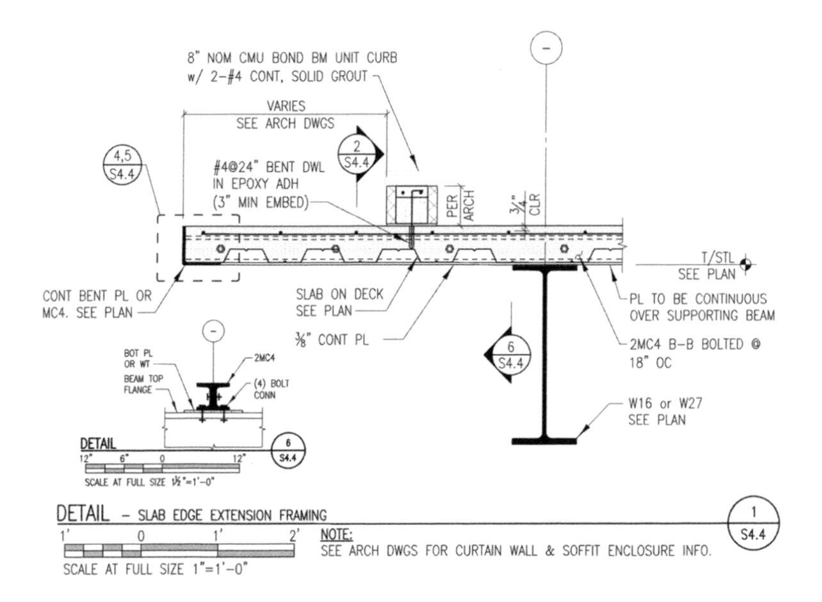

Figure 1.5 Slab Extension

Source: Keast & Hood

seismic area, we decided to join them with drag struts at the 5th and 6th floors.

With the potential of large temperature swings during construction, we were also concerned that thermal forces might overstress the drag struts. If restrained against shrinkage or expansion, steel and concrete can overwhelm **connections** to supports. Bridge designers know this well and therefore connect one end of bridges on roller supports so the structure can "breathe" with seasonal temperature changes.

Façade elements were another challenge. Panel weights and restrictive **deflection** limitations forced the spandrel **beams** to be very heavy. The architectural precast wall panels were designed as 8-foot (2.44 m) wide, 2-story tall elements imposing their weight eccentrically on spandrel beams. We designed rotation counters perpendicular to these spandrel beams but our assumed **eccentricity** of 6 in (152 mm) to the precast load was not enough. When the **shop drawings** arrived, we saw they had between 8 and 10 in (203 and 254 mm) of eccentricity and the loads did not align well with the rotational restraint of the beams.

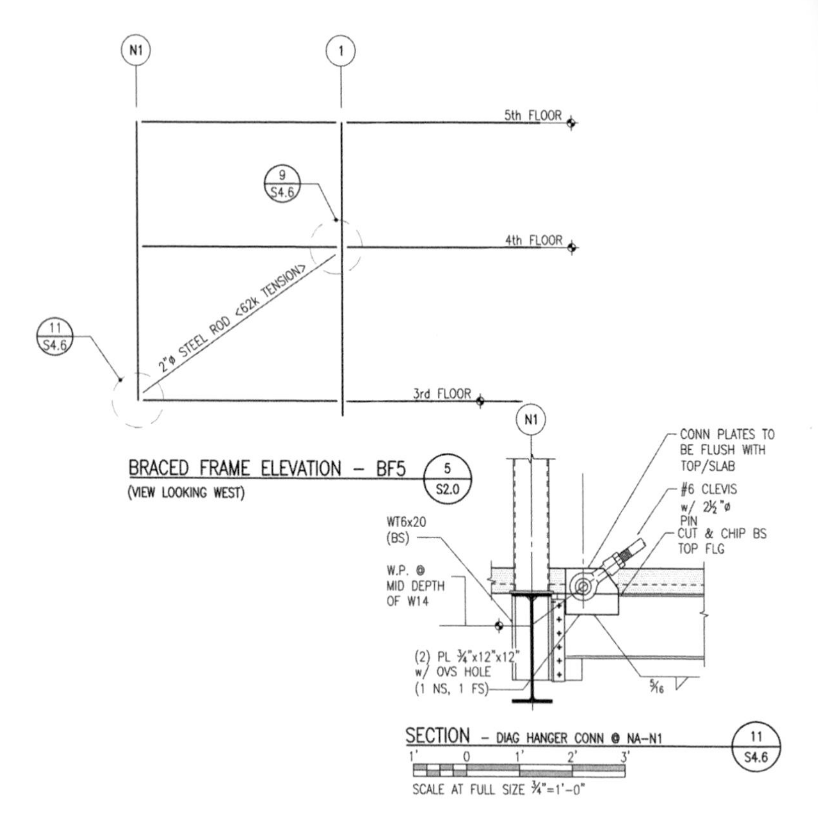

Figure 1.6 N-W Corner Diagonal Brace

Source: Keast & Hood

In hindsight, we might have been better off with structural tubes as spandrel beams or plating one side of the **wide flange** members to add **torsional** strength and rigidity. The fix was field installation of diagonal kickers at precast connection locations not backed up by a torsional restraint beam.

Another challenge we faced was with the glass curtain wall contractor who wanted to "panelize" his system to expedite erection and decrease labor costs. The architecture did not permit topside connections to the slab so the connections were hidden below. This under-slab connection suggested by the envelope consultant and designed by the curtain wall supplier's engineer required close coordination with the steel spandrel beams and a watchful eye in the field during construction.

Information flow, critical to any large project, has to be open and honest. Facilitating this throughout the various project stages was an excellent project architect. Keeping everybody on board and working toward a common goal was his top priority.

Coordination between the consultants is of vital importance, especially when the structural design precedes the mechanical system selection. At a lecture on the Drexel campus in the 1990s, the author heard Leslie Robertson, the noted structural engineer, say "We anticipate and make allowances for the mechanical systems". He was referring to allowances they make for uniformly distributed weights of piping, ductwork, etc., and the requisite vertical shaft space, not major pieces of equipment. For heavy rooftop equipment, we relied on the MEP consultant's help to estimate the weight of large rooftop equipment even though the final weights were unknown until the equipment was purchased.

In the case of boiler rooms, suspended piping is large and dense; therefore, a much greater uniform load is suspended from the structure above. We typically assume a uniformly distributed 25 lb/ft^2 (1.2 kN/m^2) suspended load in "engine" rooms. This must be verified with actual equipment and piping after the mechanical design is complete.

Elevator equipment geometry and loading depends on the elevator supplier, the machine type, travel speed, capacity, and other variables. We were fortunate to have input from an elevator consultant hired by the architect to define minimum shaft size, pit dimensions, loads, etc. This information was checked against the vendor's submittals much later in the project, after the steel was designed and foundation work had begun.

1.3 CONSTRUCTION DOCUMENTS

Notice to proceed was given in January of 2011, with a May 2011 due date for final **construction documents**. Many of the questions raised at the end of the DD phase remained. As design progressed and we incorporated the accepted VE changes, other costs crept back into the project. Nothing was off the table in the search for savings. What more could we do structurally?

We worked long hours to interpret architectural intent, refine the computer models, and develop the final design. The rod-suspended monumental stair alone consumed over two hundred hours of design and drawing effort. We modeled it twice, once with unyielding supports to maximize the hanger rod force and a second time replacing the unyielding supports with springs predicted based on the RAM model with a unit load on the steel cantilevers to represent the building steel in its deflected position. Until late in the design, we did not know whether the university would fund one or three levels of this design element.

The story high trusses and monumental stair were analyzed in STAAD®, while the rest of the steel structure was analyzed in RAM Structural System®. The foundation mat was designed using RAM Concept®. Many hand calculations filled in the blanks where the computer analysis was not suitable or was limited.

We "fooled" the software by adding story-high diagonal bracing in place of the trusses. From this model, the estimated story shear transfer force caused by wind or seismic was manually added to the web forces and increased the web sizes.

There were delays and a nagging lack of information on several key elements:

- Precast wall panel support
- Aluminum curtain wall support
- Monumental stair geometry and number of levels
- Rooftop mechanical equipment
- Floor box locations, fireproofing, etc.
- Slab edge dimensions and openings
- For the large atrium smoke evacuation ductwork needed, we cleared a path below and through the roof structure.

1.4 CONSTRUCTION ADMINISTRATION

We headed into construction with many unresolved issues and missing information on several key mechanical systems, but felt that we were on solid ground with respect to the architecture we were supporting.

During construction, a system of steel rakers (Figure 1.7) adjacent to the subway braced the soldier piles and penetrated the positive, "blind-side" waterproofing attached to the site retention system with the one-sided formwork attached to the same vertical system. To resist the lateral loads, there were thrust blocks cast beneath the mat. Also, an architectural issue/concern was the waterproofing details around these penetrations below the mat foundation.

Figure 1.7 Rakers and Mud Mat

A conventionally built site retention system surrounded the other sides of the excavation. Tied with soil anchors or tiebacks, these and the soldier piles were sacrificial, in other words, they will remain buried and forgotten.

The transfer trusses spanning above the future auditorium were "stick-built", meaning they were bolted up in the field. Connections utilized **slip critical** bolts because the truss member forces would reverse after temporary column removal (refer to Figure 1.3).

The **AISC Code** of Standard Practice encourages the EOR to communicate any "Special erection limitations". Decommissioning the temporary columns dictated the load transfer from these to the trusses; therefore, we proposed a sequence for their removal on the documents. The fabricator asked several questions about the trusses including the following:

- What amount of **camber** did we want?—answer: none because the deflection we predicted would be minimal.

Figure 1.8 Transfer Truss

- What was the estimated load carried by the temporary columns?—answer: about 300 k (1335 kN).
- When could the temporary columns be removed?—answer: 7 days after the third-floor slab was cast so the top chord would be laterally braced by the floor slab.

After the temporary columns were removed the deflection measured in the field was less than 1/2 in (12.7 mm).

Specialty system designs for **components** of the building such as steel stairs, light-gage steel studs for interior and exterior architectural finishes, precast architectural wall panels, and aluminum curtain wall systems were delegated to the supplier's registered engineer. During construction, we resolved conflicts regarding assumed support points and building structure limitations.

We learned several valuable lessons regarding the façade. When supporting systems designed by others, engage the services of the component suppliers' engineers or a consultant knowledgeable about those systems and get preliminary design information. Façade systems are very expensive building components and the construction manager will unavoidably select the lowest bidder, and often these subcontractors have their own concepts for how their systems will be supported. We discovered that the precast supplier did not understand the specifications regarding connection design requirements and we painfully resolved these last-minute coordination issues during construction.

Around the rooftop equipment well was a 16'-tall screen-wall of precast concrete panels to hide the mechanical systems. To avoid large deflections, the 14'-tall, cantilevered columns were cover-plated and continued without a **splice** down to the next floor. Because these were outside, the plates were "seal-welded" to avoid hidden crevices where the galvanizing-pretreat caustics will penetrate and boil off during the hot-dip process. Late in the design, provisions were added to support window-washing equipment from this steel, but we had sufficient design **capacity** to carry the load.

Figure 1.9 Fine-Tuning the Hanger Rods

Another component that required buy-in from the supplier's engineer was the cantilevered and suspended monumental stair. We (the EOR) modeled the stair so we could design the supporting structure. The stair was highly redundant with multiple load pathways back to the main structure. During construction, to avoid overloading some parts of the structure and underutilizing others, we requested help from the university's engineering faculty in "tuning" the hangers. This involved applying strain gages to the hanger rods prior to erection. Once erected, the gages were connected to a data capture system and the steel workers adjusted the **tension** until it was within range of our assumptions.

Testing and inspection is often a "hot-potato" issue because the owner pays these unknown costs throughout construction. Even with the unknowns, these services are estimated based on construction schedule and project complexity, and therefore the lowest bidder usually gets the work. Be careful how you craft the RFP for inspection services so you don't leave anything out. For LeBow, a late scope addition was for instrumenting the stair hanger rod tension.

Figure 1.10 Monumental Stair Looking Down
Source: Keast & Hood

Figure 1.11 East Façade Cantilever
Source: Allison Worthington, Keast & Hood

Figure 1.12 East Façade
Source: Allison Worthington, Keast & Hood

1.5 OPENING

According to the university, the students love their new building; most take the stairs instead of the elevators. The school is considered one of the best worldwide and in the top 10% nationwide.

NOTE

1 "GeoHistory Resources," Greater Philadelphia GeoHistory Network, accessed May 1, 2017. www.philageohistory.org/rdic-images/index2.cfm

Steel Fundamentals

Chapter 2

Paul W. McMullin

Structural steel is the backbone of much of today's infrastructure. Formed into beams, columns, and braces, it soars in the world's tallest buildings and longest bridges. It is found in warehouses, industrial facilities, libraries, schools, office buildings, and homes. Its strength and stiffness are unmatched by other structural materials. This chapter introduces fundamental steel design principles that will support the remaining chapters of this book.

2.1 HISTORICAL OVERVIEW

Until the late 19th century metals were used sparingly in building construction. The Greeks and Romans used bronze and iron cramps to join stone blocks located in walls with higher stresses. The builders of the late Gothic and Renaissance periods used cramps and tie rods to resist the outward thrust of vaulted ceilings and roofs.

Widespread use of iron—and eventually steel—did not occur until the mid-1800s. It closely followed the technological advancements of the Industrial Revolution. During the population and economic boom of the late

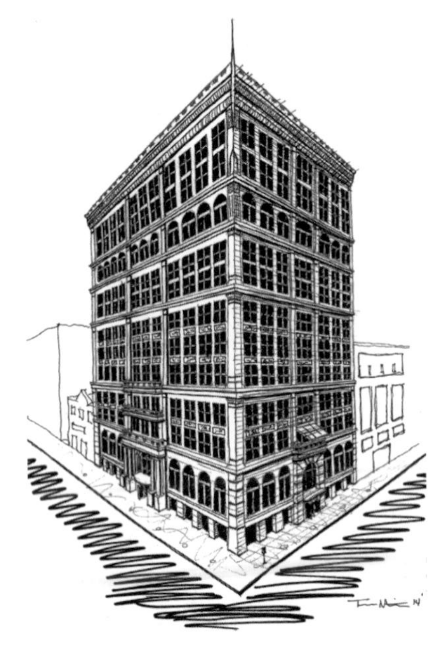

Figure 2.1 Home Insurance Building, William LeBaron Jenney, 1893, Chicago, USA

19th and early 20th centuries American cities grew rapidly. Larger and less costly structures were needed to house the expanding population. Steel filled this need.

The Great Chicago Fire of 1871 gave American architects the opportunity to leave behind the historicism of Europe and create a distinctly American architectural style—the skyscraper (Stokstadt 2002, 1054). The Home Insurance Building, completed in 1884 in Chicago by architect William LeBaron Jenney, was the first tall building with a skeletal **structural system** independent from its exterior "skin" wall, illustrated in Figure 2.1.

Modernism, fueled by new technologies and materials, helped drive the trend towards new architectural forms and aesthetics. The International Style became popular for skyscrapers and high-rises. By the mid-1970s, the modernist aesthetic preferences with its simplified, geometric forms, and its clean, unornamented facades lost ground to the reactionary architecture of Postmodernism. Postmodernism was supplanted by High-Tech architecture, Neo-modernism, and the sustainable architectural

Figure 2.2 Burj Khalifa, SOM, 2004–2010, Dubai, UAE

innovations of today. The Burj Khalifa (Figure 2.2), completed in 2010 by SOM architects, is an example of the structural liberation that is achievable with skeletal steel construction.[1]

2.2 CODES

The American Institute of Steel Construction (AISC) and a few companion organizations maintain the primary codes we used to design structural steel in the United States. Similar organizations exist in other countries, with similar codes. The following are the most common codes, and a brief description is included.

AISC 360 Specification for Structural Steel Buildings[2] contains the fundamental provisions for structural steel buildings. If you wish to further investigate a specific topic, Table 2.1 lists its contents.

AISC 303 Code of Standard Practice for Steel Buildings and Bridges[3] contains trade practices for the fabrication and erection of structural steel buildings and bridges.

Specification for Structural Joints Using High-Strength Joints[4] provides detailed requirements for bolted connections.

AISC 341 Seismic Provisions for Structural Steel Buildings[5] guides engineers through the many seismic design requirements, many of which are a produce of the 1994 Northridge earthquake.

AISC combines the first three specifications into the *Steel Construction* Manual,[6] along with a substantial set of design guidance and tables, listed in Table 2.2.

AISC also publishes a series of Design Guides, on topics ranging from base plates to staggered trusses to sound isolation. They are most useful when we need more detailed information on a specific topic.

2.1.1 Strength Reduction ϕ Factors

Safety factors reduce the chance of failure. In **strength design** (**LRFD**), we apply these to both **demand** and capacity (load and strength). Figure 2.3 illustrates the relationship between demand and capacity. **Load factors** on **load combinations** push the demand curve to the right, while **strength reduction factors** ϕ shift the capacity curve to the left. Proportioning members is based on providing greater capacity than demand. Where the curves overlap, failure can occur. Even if the structure is designed without errors, statistically, there is a slight (0.01%) chance of failure. The curve shape in Figure 2.3 is a function of

Table 2.1 AISC 360 code summary

Section	*Title*	*Contents*
Chapters		
A	General Provisions	Scope, referenced specifications, materials, and drawing requirements
B	Design Requirements	Loads, design basis, member properties, local buckling, fabrication, erection, quality, existing structures
C	Stability	General stability requirements
D	Tension	Tension member design
E	Compression	Compression member design
F	Flexure	Flexure (bending) member design
G	Shear	Shear member design
H	Combined forces	Design for combined tension, bending, compression, and torsion
I	Composite members	Requirements for composite steel and concrete members
J	Connections	Provisions for welds, bolts, and connected elements
K	HSS and Box Section Connections	Additional requirements for HSS and box member connections
L	Serviceability	Requirements for deflection, drift, vibration, wind, thermal change, and connection slip
M	Fabrication and Erection	Shop and erection drawings, fabrication, painting, and erection
N	Quality	Quality control and assurance requirements
Appendices		
1	Advanced Analysis	Requirements for elastic and inelastic analysis
2	Ponding	Ponding design requirements
3	Fatigue	Fatigue design provisions
4	Fire	Structural design for fire conditions

Table 2.1 *continued*

Section	*Title*	*Contents*
5	Existing Structures	Evaluation criteria for existing structures
6	Stability Bracing	Column, beam, and beam-column bracing
7	Alternate Stability	Alternate stability design methods
8	Second-Order Analysis	Approximate methods for second order analysis

The commentary follows the chapters and appendices order above

variability. Narrow curves have less deviation from average than wider curves.

You might think it would make sense to provide substantially greater capacity than demand. This would be true if no one had to pay for the materials, or our environment didn't have to support extraction and manufacturing costs. Since both are of concern, we balance demand with capacity to minimize risk, cost burdens, schedule, and environmental impacts. This is the art of engineering.

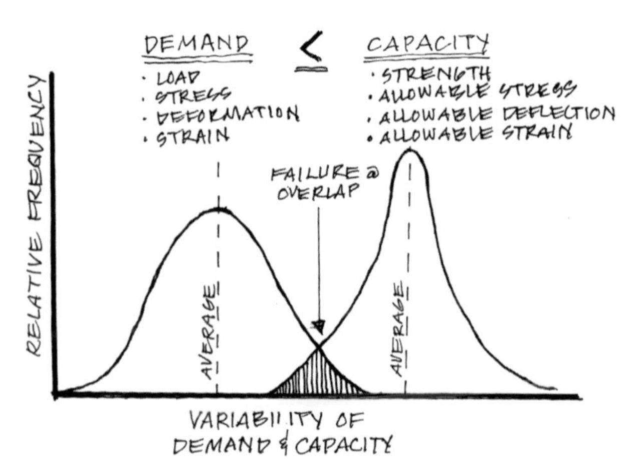

Figure 2.3 Strength and load distribution relationship

Table 2.2 Steel Construction Manual section summary

Part	*Title*	*Contents*
1	Dimensions and Properties	Dimensions and section properties for all shapes
2	General Design Considerations	Basic code, safety, and design guidance
3	Flexural Members	Design tables for flexure (bending)
4	Compression Members	Design tables for compression
5	Tension Members	Design tables for tension
6	Combined Forces	Design tables for combined member forces
7	Bolts	Tables for bolt strength and geometry
8	Welds	Prequalified welds, and strength tables
9	Connecting Elements	Hole reduction, block shear for wide flange webs
10	Shear Connections	Double angle, end plate, seated, single plate, single angle, and tee connection tables
11	PR Moment Connections	Provisions for partially restrained moment connections
12	FR Moment Connections	Provisions for fully restrained moment connections
13	Brace and Truss Connections	Provisions for brace and truss connections
14	Bearing and Column Connections	Guidance for beam bearing plates, column base plates, anchor rods, and column splices
15	Special Connections	Criteria for hanger, bracket, and crane rail connections
16	Specification	AISC 360 specification
17	Miscellaneous	Metric conversions and equivalent shapes, weights

Strength reduction factors, or phi ϕ factors, account for variations in material quality and construction tolerances. Statistically derived from testing, these are less than one (reducing strength); with lower factors indicating greater variation. They reduce the nominal steel design strength to ensure safe performance. Table 2.3 summarizes strength reduction factors for steel design. Consult each chapter for the exact factor.

2.3 MATERIALS

Many structural steels are available today for use in steel construction. *ASTM International* writes the material specifications that govern their properties. Other countries have similar specification bodies, such as the BSI Group in the UK and the JIS in Japan.

There is usually one type of steel that is the most common for a given structural shape. For instance, A992 is the most common specification for wide-flange shapes in the United States. However, it is possible to get wide flanges made of different steel specifications, such as A572. Table 2.4 and Table 2.5 show the most common structural steel specifications, strength values, and their availability for a given shape group and plate, respectively.

Looking closer at Table 2.4 we see it is grouped into three sections. Exploring these further,

- **Carbon Steels**—Iron and carbon, where the carbon provides the primary alloying element.
- **High-Strength, Low-Alloy Steels**—Iron alloyed with elements like copper, nickel, vanadium, and chromium, providing greater mechanical properties.

Table 2.3 Phi factors for various steel strength modes

Strength Mode	ϕ
Tension	0.9
Bending	0.9
Shear	0.9–1.0
Compression	0.9
Connections	0.65–1.0

Source: AISC 360–16

- **Weathering Steels**—Iron alloyed with a variety of elements to form a stable oxide on the surface that prevents further corrosion (like aluminum).

In addition to structural shapes and plate, manufacturers produce many fasteners. Table 2.6 provides a list of common fastener types. These will be discussed at greater length in Chapter 8.

2.4 MATERIAL BEHAVIOR

2.4.1 Stress and Strain

The **stress–strain** behavior of steel consists of two primary regions: **elastic** and **inelastic**, as illustrated in Figure 2.4.

When we remove load in the elastic range, the material returns to its original shape, like the paperclip we don't bend very far. If we venture into the inelastic range the material will be permanently deformed when load is removed—like when you really crank on the paperclip. Actual stress–strain follows the dashed curve in Figure 2.4. We simplify the curve by assuming the steel stress is constant after **yield**.

The area under the stress-strain curve is the energy the material can absorb, illustrated in Figure 2.5 for a low- and high-deformation material. Steel follows the high deformation curve. In situations where we want the structure to dissipate energy, such as during earthquakes and blast events, using a material with high strain potential after yield is fundamental.

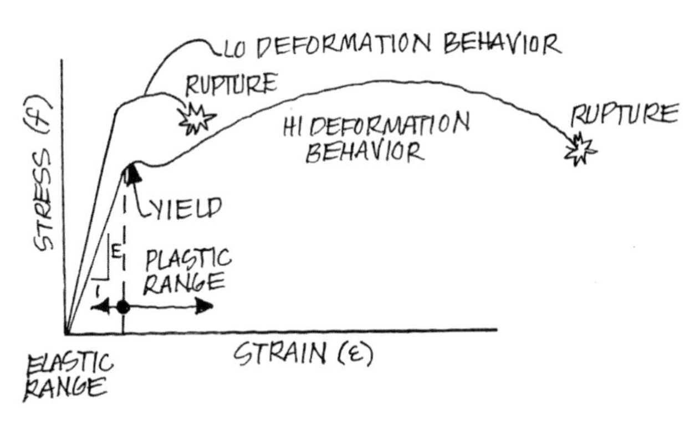

Figure 2.4 Stress–strain relationship for structural steel

Table 2.4 **Representative structural steel shape grades and their availability**

			Imperial Strength		*Metric Strength*		*Shape Series*								
Steel Type	*ASTM Designation*	*Grade*	F_y *(k/in²)*	F_u *(k/in²)*	F_y *(MN/m²)*	F_u *(MN/m²)*	*Wide Flange (W)*	*Standard (S)*	*Piles (HP)*	*Channels (C)*	*Miscellaneous Channels (MC)*	*Angle (L)*	*Rectangular HSS*	*Round HSS*	*Pipe*
Carbon	A36		36	58	250	400									
	A53	B	35	60	241	414									
	A500	B	42	58	290	400									
		B	46	58	317	400									
	A501	A	36	58	250	400									
	A529	50	50	65	345	448									
High-Strength Low-Alloy	A572	42	42	60	290	414									
		50	50	65	345	448									
	A618	I,&II	50	70	345	483									
	A913	50	50	60	345	414									
	A992		50	65	345	448									

Corrosion Resistant	A242		42	63	290	434									
			50	70	345	483									
	A588		50	70	345	483									
	A847		50	70	345	483									

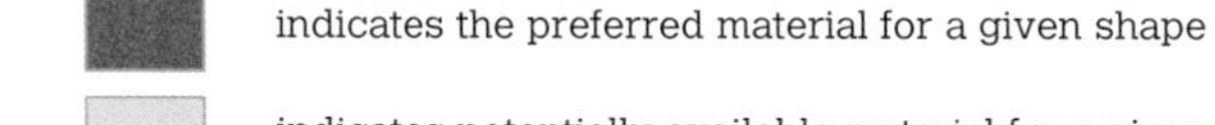

indicates the preferred material for a given shape

indicates potentially available material for a given shape, verify before specifying

indicates unavailable material for a given shape

Notes

1) Common and representative materials are listed. See AISC Manual for additional materials.

2) HSS= Hollow Structural Shapes.

3) Non preferred material may not be available in all flange thicknesses.

Source: AISC Manual of Steel Construction, 14th Ed.

Table 2.5 Representative structural steel plate grades and their availability

			Imperial Strength		*Metric Strength*		*Plate Thickness (in)*									
							to 0.76	*0.76 to 1.25*	*1.21 to 1.5*	*1.1 to 2.0*	*2.6 to 2.5*	*2.6 to 4.0*	*4.1 to 5.0*	*5.1 to 6.0*	*6.1 to 8.0*	*8.1*
							Plate Thickness (mm)									
Steel Type	*ASTM Designation*	*Grade*	F_y *(k/in²)*	F_u *(k/in²)*	F_y *(MN/m²)*	F_u *(MN/m²)*	*to 19*	*20 to 32*	*33 to 38*	*39 to 51*	*52 to 64*	*65 to 102*	*103 to 127*	*128 to 152*	*153 to 203*	*204*
Carbon	A36		32	58	221	400										
			36	58	250	400										
	A529	50	50	70	345	483										
HSLA	A572	42	42	60	290	414										
		50	50	65	345	448										
Corrosion Resistant	A588	42	42	63	290	434										
		46	46	67	317	462										
		50	50	70	345	483										
Q&T	A514		90	100	621	690										
			100	110	690	758										
	A852		70	90	483	621										

indicates the preferred material for a given plate thickness

indicates potentially available material for a given plate thickness, verify before specifying

indicates unavailable material for a given plate thickness

Notes

1) Common and representative materials are listed. See AISC Manual for additional materials.

Source: AISC Manual of Steel Construction, 14th Ed.

Table 2.6 Common structural fastener types and grades

	Imperial Strength		Metric Strength		Diameter Range		Fastener Type								
ASTM Designation	F_y (k/in^2)	F_u (k/in^2)	F_y (MN/m^2)	F_u (MN/m^2)	*d (in)*	*d (mm)*	*Conventional Bolts*	*Twist-Off Bolts*	*Common Bolts*	*Nuts*	*Washers*	*Threaded Rod*	*Headed Stud Anchors*	*Headed*	*Threaded & Nutted*
A108	–	65	–	448	0.375 to 0.75	9.5 to 19							■		
A325	–	105	–	724	> 1 to 1.5	>25 to 38	■								
	–	120	–	827	0.5 to 1.0	12.7 to 25	■								
A490	–	150	–	1,034	0.5 to 1.5	12.7 to 38	■								
F1852	–	120	–	827	0.5 to 1.0	12.7 to 25		■							
F2280	–	150	–	1,034	0.5 to 1.125	12.9 to 29		■							
A194 Gr 2H	–	–	–	–	0.25 to 4.0	6.5 to 102				□					
A563	–	–	–	–	0.25 to 4.0	6.5 to 102				■					

F436	–	–	–	–	0.25 to 4.0	6.5 to 102									
A36	36	58	250	400	to 10	to 254									
A193 Gr B7	–	125	–	862	to 2.5	to 64									
A307 Gr A	–	60	–	414	0.25 to 4.0	6.5 to 102									
A572	50	65	345	448	to 4	to 102									
A588	50	70	345	483	to 4	to 102									
F1554	36	58	250	400	0.25 to 4.0	6.5 to 102									
	55	75	379	517	0.25 to 4.0	6.5 to 102									
	105	125	724	862	0.25 to 3.0	6.5 to 76									

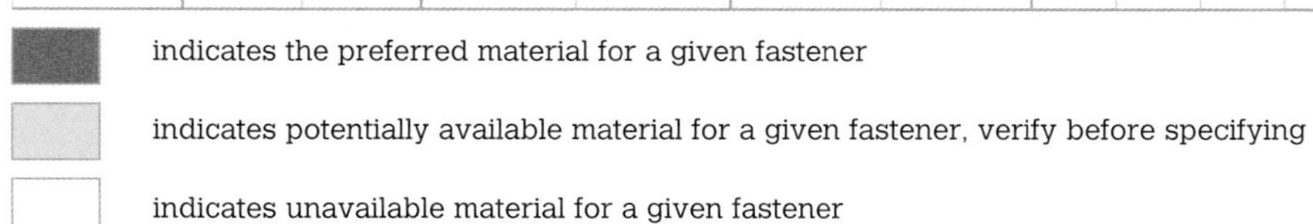

indicates the preferred material for a given fastener

indicates potentially available material for a given fastener, verify before specifying

indicates unavailable material for a given fastener

Notes 1) Common and representative materials are listed. See AISC Manual for additional materials.

Source: AISC Manual of Steel Construction, 14th Ed.

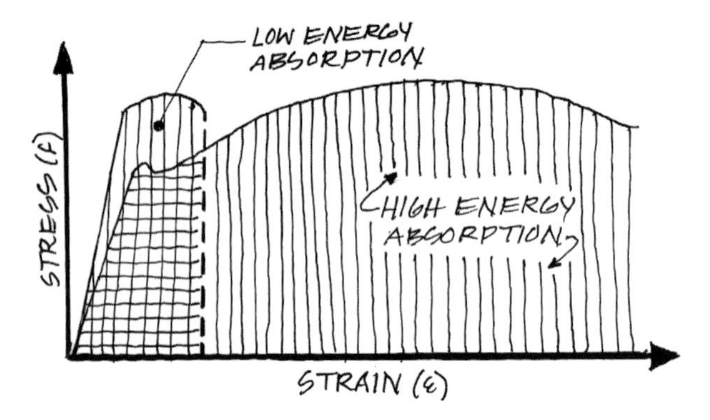

Figure 2.5 Energy absorption of low- and high-deformation materials

The previous stress–strain curves are known as the engineering type, where we find stress by dividing the force by the initial cross-section **area**. For materials that yield significantly during loading, the area reduces as the member necks down. To capture this effect, we use a true stress–strain curve, which divides load by the actual area at that load. This generates the curve shown in Figure 2.6, which shows the actual stress in a member continues to increase until failure. Note when the stress–strain curve is

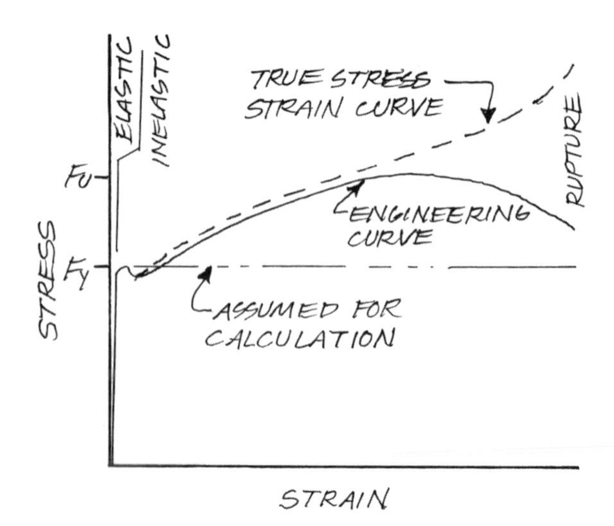

Figure 2.6 True and engineering stress–strain curve

drawn to scale, like in Figure 2.6, how small the elastic range is compared to inelastic. To simplify things, we typically assume the yield stress is constant until it reaches the **rupture** strength.

2.4.2 Local Buckling

Because steel is so strong, compared to its structural material counterparts, it takes much less of it to carry loads. This results in structural shapes that are prone to **local buckling**, illustrated in a soda can in Figure 2.7. Local buckling is like global buckling, except only a portion of the member deforms, illustrated in Figure 2.8. As you can imagine, this has the potential to substantially degrade the member's strength.

Propensity for local buckling is a function of the edge restraints, yield strength, and **modulus of elasticity**. The more edge restraint—like the web of a **channel**—the less likely the section is to locally buckle. This produces types of local buckling, unstiffened and stiffened. An unstiffened element has one edge (parallel to the stress) supported and one edge unsupported, while a stiffened element has both edges supported, illustrated in Figure 2.9.

To begin, we need to understand the geometry that defines local buckling. We compare the length to **width** ratio of a local element to

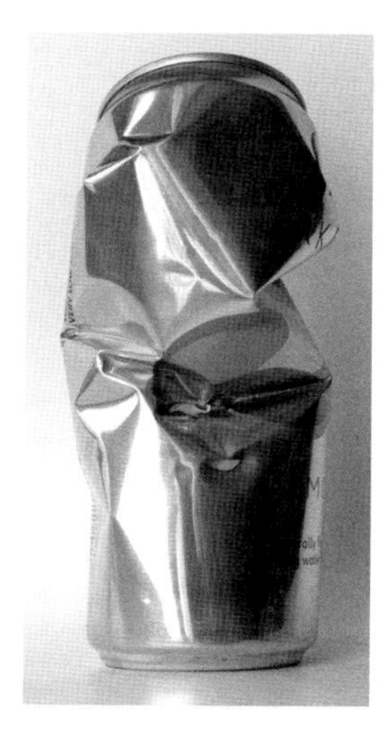

Figure 2.7 Local buckling in a soda can

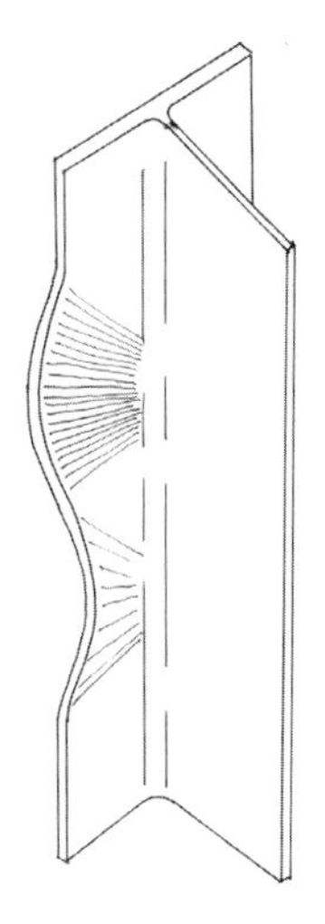

Figure 2.8 Local buckling in a wide flange, angle, and pipe

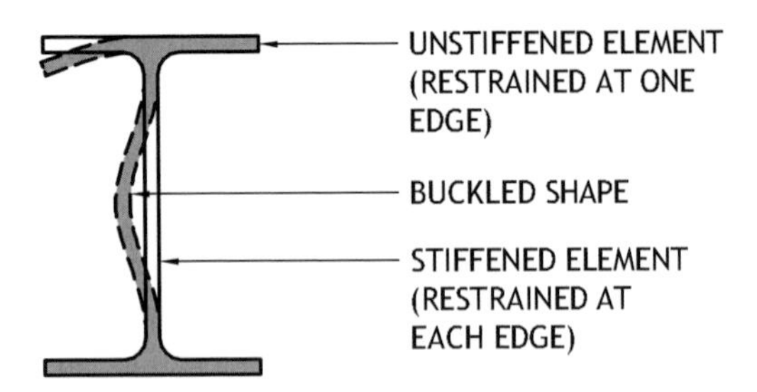

Figure 2.9 Definition of stiffened and unstiffened elements

limiting values. For a wide flange, we look at the flange width to thickness ratio (b/t) and web length to thickness ratio (h/t_w). These dimensions are defined in Figure 2.10. We then compare these ratios to the limiting width-to-thickness ratio which is a function of the square root of E/F_y. The nonslender to **slender** transition for compression λ_r defines the point where the element becomes slender and will locally buckle in compression, listed in Table 2.7.

For flexure, the code defines two limits, relating to the amount of inelastic buckling. **Compact** shapes will fully yield before local buckling, and are defined by the compact limit λ_p. Non-compact shapes will locally buckle in the inelastic range (meaning they locally yield) between the compact and slender limits. Shapes beyond the slender limit λ_r , will elastically buckle and come back to their original shape if the load is removed. Table 2.8 provides flexural buckling width-to-thickness limits.

The section property tables in the *AISC Steel Manual* identify when a shape is locally slender, or non-compact.

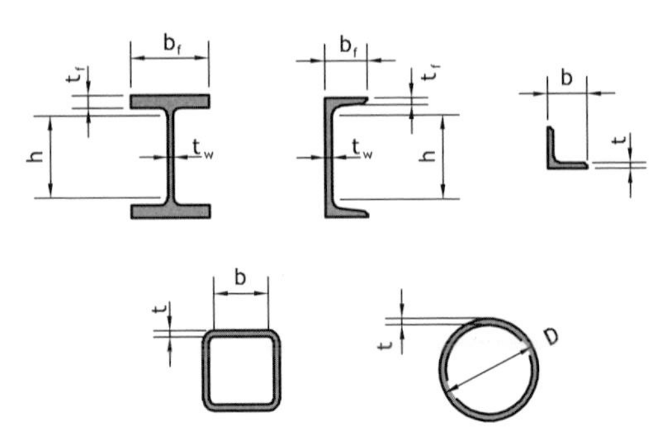

Figure 2.10 Local buckling dimension definitions for common shapes

Table 2.7 Limiting width-to-thickness ratios for select compression elements

	Element	*Width/ Thickness Ratio*	*Compression Limiting Ratio*[1] λ_r	*Figure*
Unstiffened Elements	Flanges of rolled I-shaped sections, channels, and tees	b/t	$0.56\sqrt{E/F_y}$	
	Flanges of built up I-shaped sections	b/t	$0.64\sqrt{E/F_y}$	
	Legs of angles	b/t	$0.45\sqrt{E/F_y}$	
	Stems of tees	d/t	$0.75\sqrt{E/F_y}$	
Stiffened Elements	Webs of doubly symmetric I-shaped sections and channels	h/t_w	$1.49\sqrt{E/F_y}$	
	Walls of rectangular HSS and box sections	b/t	$1.40\sqrt{E/F_y}$	
	Other stiffened elements	b/t	$1.49\sqrt{E/F_y}$	
	Round HSS	D/t	$0.11E/F_y$	

Note
1) Nonslender to slender transition
Source: AISC 360–16

Local buckling reduces member strength. The following chapters primarily focus on compact sections, so we can ignore the effects of local buckling. However, when a section is slender, or non-compact, the equations governing bending, **shear**, and compression strength will change.

2.4.2 Fracture Toughness

Toughness is to the boxer, as strength is to the weight lifter. It is a measure of the amount of abuse a material can take. More technically, fracture toughness measures crack resistance.

Table 2.8 Limiting width-to-thickness ratios for select flexural elements

			Flexural Limiting Ratio		
	Element	*Width/ Thickness Ratio*	*Compact to Noncompact* λ_p	*Noncompact to Slender* λ_r	*Figure*
Unstiffened Elements	Flanges of rolled I-shaped sections, channels, and tees	b/t	$0.38\sqrt{E/F_y}$	$1.0\sqrt{E/F_y}$	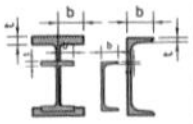
	Flanges of built up I-shaped sections	b/t	$0.38\sqrt{E/F_y}$	$0.95\sqrt{E/F_y}$	
	Legs of angles	b/t	$0.54\sqrt{E/F_y}$	$0.91\sqrt{E/F_y}$	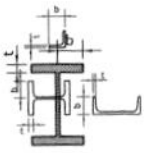
	Flanges of I-shaped sections and channels in flexure about their weak axis	b/t	$0.38\sqrt{E/F_y}$	$1.0\sqrt{E/F_y}$	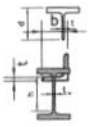
	Stems of tees	d/t	$0.84\sqrt{E/F_y}$	$1.52\sqrt{E/F_y}$	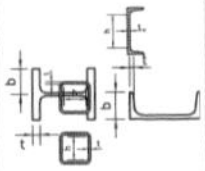
Stiffened Elements	Webs of doubly symmetric I-shaped sections and channels	h/t_w	$3.76\sqrt{E/F_y}$	$5.7\sqrt{E/F_y}$	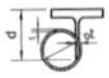
	Flanges of rectangular HSS and box sections	b/t	$1.12\sqrt{E/F_y}$	$1.4\sqrt{E/F_y}$	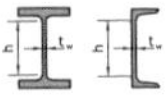
	Walls of rectangular HSS and box sections	b/t	$2.42\sqrt{E/F_y}$	$5.7\sqrt{E/F_y}$	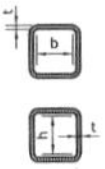
	Round HSS	D/t	$0.07E/F_y$	$0.31E/F_y$	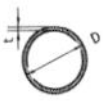

Source: AISC 360–16

Fracture toughness and all structural properties are influenced by size. As size increases, toughness—and strength, stiffness, and stability—decrease, as shown in Figure 2.11. As we move from a test specimen to component test to full-scale structure, there is a general trend of decreasing material properties.

Specific to fracture toughness, as the material gets thicker, the toughness decreases, shown in Figure 2.12. However, after a certain point, the toughness remains the same—known as the **plane strain** fracture toughness.

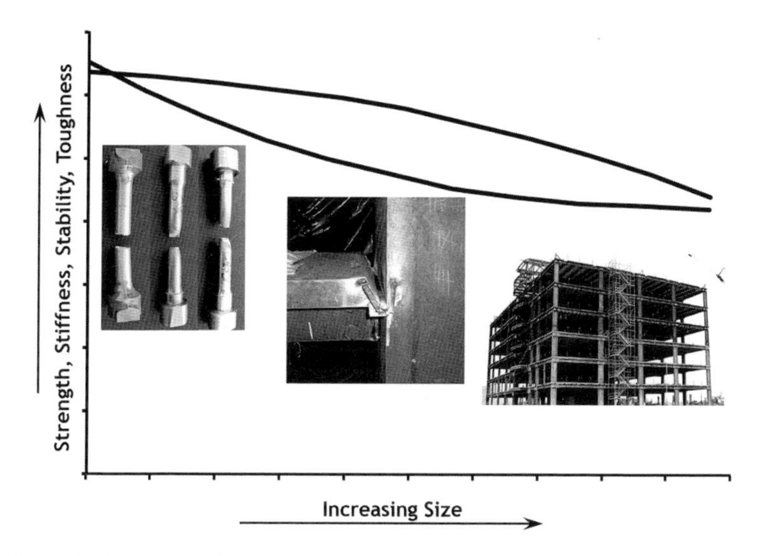

Figure 2.11 Conceptual material property variation with change in size

Photos courtesy Bill Komlos

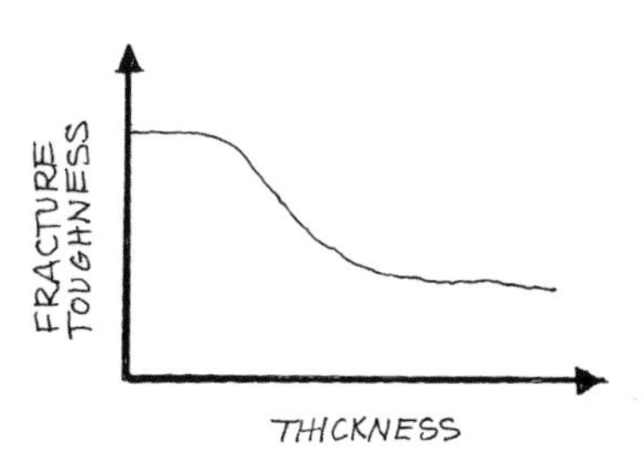

Figure 2.12 Fracture toughness as a function of thickness

2.5 STRUCTURAL CONFIGURATION

Structural steel is common in simple structures, but fabrication and erection techniques make it possible in highly complex and ornate structures. Similarly, it is often hidden behind architectural finishes, but can dramatically stand on its own, as we see in Figure 2.13.

Common steel framing consists of columns on a regular grid, a concrete slab over metal deck, with beams supporting the slab. Often in commercial structures, the steel beams have headed studs welded to them to engage the concrete slab, seen in Figure 2.14. The system then acts like a reinforced concrete beam, and reduces the beam weight.

To aid in the initial layout of a structure the following tables provide:

- Typical plan dimensions and floor to floor heights—Table 2.9.
- Estimated depths for different horizontal members and spans—Table 2.10.
- Estimated column section sizes with varying **tributary areas**—Table 2.11.

Figure 2.13 Real Salt Lake Soccer Stadium

Photo courtesy of Mark Steel Corporation

Figure 2.14 Headed studs welded to a floor beam prior to concrete slab placement

Table 2.9 Typical plan dimensions and floor to floor heights

	Bay Plan Dimensions		
Building Type	*Short ft (m)*	*Long ft (m)*	*Floor to Floor ft (m)*
Hospital	25–35	30–40	15–20
	(7.5–10.5)	(9–12)	(4.5–6)
Hotel	15–30	25–35	12–15
	(4.5–9)	(7.5–10.5)	(3.5–4.5)
Office	25–35	30–50	13–18
	(7.5–10.5)	(9–15)	(4–5.5)
Parking	18–27	30–60	12–15
	(5.5–8.5)	(9–18)	(3.5–4.5)
Warehouse	20–40	35–50	18–30
	(6–12)	(10.5–15)	(5.5–9)

Table 2.10 Estimated depths for different horizontal members and spans

Imperial Measures			*Span (ft)*											
		Span/ Depth	*10*	*15*	*20*	*25*	*30*	*40*	*50*	*75*	*100*	*150*	*200*	*300*
System			*Depth (in)*											
Beam		**20**		9	12	15	18	24	30	45				
Composite Beam		**28**		8	9	11	13	17	21	32				
Crane Girder		**10**		18	24	30	36	48	60	90				
Floor Joist		**20**	6	9	12	15	18	24	30	45				
Roof Joist		**24**	6	8	10	13	15	20	25	38	50			
Plate Girder		**15**			16	20	24	32	40	60	80			
Truss		**12**			20	25	30	40	50	75	100	150	200	300
Space Frame		**16**					23	30	38	56	75	113	150	225

SPAN DEPTH

Metric Measures		*Span (m)*											
	Span/ Depth	*3*	*4.5*	*6*	*7.5*	*9*	*12*	*15*	*23*	*30*	*45*	*60*	*90*
		Depth (mm)											
Beam	**20**		230	300	390	460	610	770	1150				
Composite Beam	**28**		200	220	270	320	440	550	820				
Crane Girder	**10**		460	610	760	900	1200	1500	2300				
Floor Joist	**20**	150	230	300	400	450	600	750	1150				
Roof Joist	**24**	150	200	250	320	380	500	640	950	1250			
Plate Girder	**15**			400	500	600	800	1000	1500	2000			
Truss	**12**			500	640	760	1000	1300	1900	2500	3800	5000	7600
Space Frame	16					570	760	950	1400	1900	2860	3800	5700

Notes

1) This table is for preliminary sizing only. Final section sizes must be calculated based on actual loading, length, and section size

2) Span ranges indicated are typical. Longer spans can be made with special consideration

3) PT= post-tensioned

Table 2.11 Estimated column section sizes for given number of stories and tributary area

Tributary Area (ft^2)	*Imperial Shapes*		Number Stories for Approximately 25'–0 (7.5 m) Square Bay	*Tributary Area (m^2)*	*Metric Shapes*	
250	W4 × 13	W100 × 19.3		25	HSS4 × 4 × 3/16	HSS101.6 × 101.6 × 4.8
500	W6 × 16	W150 × 24.0	1	35	HSS4 × 4 × 1/4	HSS101.6 × 101.6 × 6.4
1,000	W6 × 20	W150 × 29.8	2	90	HSS4 × 4 × 3/8	HSS101.6 × 101.6 × 9.5
	W8 × 21	W200 × 31.3			HSS5 × 5 × 1/4	HSS127 × 127 × 6.4
1,500	W8 × 24	W200 × 35.9	3	140	HSS5 × 5 × 3/8	HSS127 × 127 × 9.5
	W10 × 30	W250 × 44.8			HSS6 × 6 × 1/4	HSS152.4 × 152.4 × 6.4
2,000	W8 × 31	W200 × 46.1	4	190	HSS5 × 5 × 1/2	HSS127 × 127 × 12.7
	W10 × 33	W250 × 49.1			HSS6 × 6 × 3/8	HSS152.4 × 152.4 × 9.5
3,000	W8 × 40	W200 × 59	5	280	HSS6 × 6 × 5/8	HSS152.4 × 152.4 × 15.9
	W10 × 39	W250 × 58			HSS8 × 8 × 3/8	HSS203.2 × 203.2 × 9.5
4,000	W8 × 67	W200 × 100	7	370	HSS8 × 8 × 1/2	HSS203.2 × 203.2 × 12.7
	W10 × 49	W250 × 73				
5,000	W10 × 54	W250 × 80	8	470	HSS8 × 8 × 5/8	HSS203.2 × 203.2 × 15.9
	W12 × 58	W310 × 86			HSS10 × 10 × 1/2	HSS254 × 254 × 12.7
6,000	W10 × 77	W250 × 115	10	560	HSS10 × 10 × 5/8	HSS254 × 254 × 15.9
	W12 × 72	W310 × 107				

8,000	W10 × 100	W250 × 149	13	750		
	W12 × 87	W310 × 129			HSS12 × 12 × 1/2	HSS304.8 × 304.8 × 12.7
10,000	W10 × 112	W250 × 167	16	900	HSS12 × 12 × 5/8	HSS304.8 × 304.8 × 15.9
	W12 × 106	W310 × 158			HSS14 × 14 × 3/8	HSS355.6 × 355.6 × 15.9
	W14 × 99	W360 × 147				
20,000	W12 × 230	W310 × 342	32	1800	W14 × 257	W360 × 382
40,000	W14 × 398	W360 × 592	65	3700		
60,000	W14 × 550	W360 × 818	100	5600		
90,000	W14 × 730	W360 × 1086	150	8500		

Notes

1) This table is for preliminary sizing only. Final section sizes must be calculated based on actual loading, length, and section size.

2) For normal height columns (10–12') and moderate loading.

3) For heavy loads, increase tributary area by 15%; for light loads, decrease area by 10%

4) To extend table to other shapes, match A, I_y, r_y.

As you lay out your structure, carefully consider the value gained by using a more complex structural configuration. For the right architectural feel, it is worth the money and the desired effect can be dramatic. However, for simple structures, or where the elements are mostly concealed, complexity is unjustifiable.

2.6 SECTION TYPES AND THEIR PROPERTIES

Steel design extensively uses section properties. The most common are area *A*, **plastic section modulus** *Z*, elastic **section modulus** *S*, **moment of inertia** *I*, and **radius of gyration** *r*. *A* and *Z* relate to strength, while *I* and *r* relate to stiffness and stability, respectively. *J* and *C* are torsion properties—most commonly used for closed shapes. For simple shapes, Table 2.12 provides equations for properties of round and rectangular shapes. For more complex shapes, use the tables in Appendix 1 or the *AISC Steel Construction Manual.*

The *AISC Steel Construction Manual*[7] provides dimensions and section properties for a multitude of structural shapes. Appendix A1 provides a selection of these shapes for use with this book. The tables typically have the heaviest, several intermediate, and the lightest sections in each size group.

- Wide Flange (W)—Used for beams, columns, and braces—Table A1.1
- Standard (S)—Used for crane rails
- Pile (HP)—Used for driven piles
- Channel (C)—Used for light beams and equipment—Table A1.2
- Angle (L)—Used for bracing and connections—Table A1.3
- Hollow Structural Sections (HSS)—Used for columns, beams, braces—Table A1.4
- Pipe (P)—Used for columns and bracing—Table A1.5.

2.6.1 Wide Flange Sections

Wide flange sections are the most common shape for moderate to heavy loads. They are denoted with a W, followed by their **nominal** depth and weight per foot. A W18 × 40 (W460 × 60) is exactly 17.9 in (454.7 mm) deep and 40 lb/ft (60 kg/m). Note that for heavy sections in the series, the nominal depth can be off by as much as 6 in (150 mm), which needs to be considered when laying out floor space.

Wide flange sections are grouped by flange width in the steel manual. This allows us to select shapes that are more suited as beams, columns, or

Table 2.12 Section property equations for round and rectangular shapes

Round	*Rectangle*	
r	y, x, c, h, b	
Area		
$A = \pi r^2$	$A = bh$	
Moment of Inertia		
$I = \pi r^4/4$	$I_x = bh^3/12$	$I_y = hb^3/12$
Radius of Gyration		
$r_z = r/2$	$r_x = h/\sqrt{12}$	$r_y = b/\sqrt{12}$
Plastic Section Modulus		
$Z = 8r^3/6$	$Z_x = bh^2/4$	$Z_x = bh^2/4$
Section Modulus		
$S = \pi r^3/4$	$S_x = bh^2/6$	$S_y = hb^2/6$
$Q = 2r^3/3$	$Q_x = bh^2/8$	$Q_y = bh^2/8$
at center		
Polar Moment of Inertia		
$J = \pi r^4/2$		
$J = \sqrt{I/A}$		
	for any shape	

in between. Looking at the W12 (W310) series table in Appendix A1.1, we see five groupings of shapes. The bottom two groups (W12 × 14 through W12 × 35, or W310 × 21.0 through W310 × 52 in metric) are most suited for beams. The top groups (W12 × 65 to W12 × 336, (W310 × 97 to W310 × 500)) are most suited for columns. The middle two (W12 × 40 to W12 × 58 (W310 × 60 to W310 × 86)) work well for light columns, beams with axial loads, and beams without **lateral bracing** along their length.

The W12 (W310) tables in Appendix 1.1 are typical of the organization in the steel manual. For sections in each group, the flanges will line up when spliced, an important consideration when joining columns. The other shapes in Appendix 1 have been grouped together for space reasons.

2.6.2 Channels

Channels are used for light beams, often on platforms and stairs. They make poor columns, given their narrowness in the weak direction. There are two types of channels, C and MC. Channels are designated by C followed by their actual depth by weight per foot. A C8 × 11.5 (C200 × 17.1) is exactly 8 in (203 mm) deep, and 11.5 lb/ft (17.1 kg/m). They are grouped in the steel manual by depth.

2.6.3 Angles

Angles are commonly used as light braces and connection material. They are denoted by L and followed by the actual long and short leg dimensions, followed up by the leg thickness. An L3 × 2 × 1/4 (L76 × 51× 6.4) is 3 in (76 mm) tall, 2 in (51 mm) wide, and 1/4 in (6.4 mm) thick. They are grouped in the steel manual based on leg size.

2.6.4 Hollow Structural Sections

Hollow Structural Sections are rectangular or round in shape. They are common as columns in shorter buildings, braces, beams, and truss members. They have substantially better torsional properties than their counterparts. These shapes are designated by HSS, followed their physical dimensions, and wall thickness. A rectangular HSS12 × 6 × 1/2 (HSS304.8 × 152.4 × 12.7) is 12 in (304.8 mm) deep, 6 in (152.4 mm) wide, and has a 1/2 in (12.7 mm) wall. A round HSS12.750 × 0.500 (HSS323.9 × 12.7) has a 12.75 in (323.9 mm) outside diameter and 1/2 in (12.7 mm) wall thickness. They are grouped by size.

2.6.5 Pipe

Pipes are used for light columns and braces. They are designated by nominal inside diameter—different from the actual. They come in three weights, standard (STD), extra strong (X), and double-extra strong (XX). A pipe designated as Pipe 6 X-strong (Pipe 152 X-strong) has a nominal 6 in (152 mm) inside diameter, and 5.76 in (146 mm) actual inside diameter, and 6.63 in (168 mm) actual outside diameter. They are grouped by diameter in the steel manual.

2.7 CONSTRUCTION

Steel construction progresses as follows:

- A fabricator takes the engineering drawings and prepares erection and shop drawings. The **erection drawings** show each piece of steel and where it is located, and any field details. The shop drawings show every hole, plate, and weld on each piece of steel.
- The fabricator makes each piece of steel and inspects their work. AISC approved shops have internal quality control, unapproved shops have external inspectors.
- Iron workers erect the steel, with two bolts in each connection initially and temporary bracing.
- Workers rack and plumb the structure, then final bolting and **welding** occurs, along with connections to lateral systems like shear walls.
- Workers lay down the corrugated metal deck and weld headed studs to the composite beams.
- Pump trucks place concrete over the metal deck to form the slab.

An important consideration for steel erection is whether to **field weld** or bolt most **joints**. Field welding can result in smaller joints, especially in seismic **moment connections**. However, wind can affect the shielding gas around the electrode and reduce quality. Additionally, in remote areas, field welders may be hard to find. Field bolting connections can be faster and reduce field welding, though the joints can become rather large in seismic connections. There is not one right answer, just possibilities.

2.8 QUALITY CONTROL

Quality control of steel construction is critical to its successful performance. Variability in quality may come from steel rolling, fabrication, and erection. Ensure safeguards are in place that the things that matter are inspected. Table 2.13 summarizes key inspection activities, which are

Table 2.13 Summary of steel inspection requirements

Inspection Type	*Periodic*	*Continuous*	*Requirement*
Submittals	■		Shop certifications
	■		Erection and shop drawings
	■		Mill test reports (MTR)
	■		Welder qualifications
	■		Inspector qualifications
	■		Bolt storage and installation procedures
Weld Inspection		■	Complete and partial penetration welds, except flare-bevel welds
		■	Multi-pass fillet welds
		■	Single pass fillet welds greater than 5/16 in (8 mm)
	■		Single pass fillet welds equal to or smaller than 5/16 in (8 mm)
	■		Floor and roof deck puddle welds
	■		Welding for stairs and railings
Weld Testing	■		Visually inspect all welds before releasing pieces for installation.
	■		Test all complete joint penetration welds by radiographic or ultrasonic testing.
	■		Ultrasonically test base metal thicker than 1 1/2 in (38 mm) when subject to through thickness weld strains.
	■		Magnetic particle test beam-column CJP welds
Snug Tight Bolt Inspection	■		Fastener grades are installed where indicated in drawings.
	■		Storage and cleanliness of high strength fasteners.
	■		Faying surfaces are in firm contact for snug tight bolts.
	■		Washers are installed as required.
PreTensioned and Slip Critical Bolt Inspection	■		Confirm fastener assembly and wrench systems are suitable for pretensioned and slip critical installations.
	■		Test fastener combinations and calibrate wrench systems
	■		Fasteners are in snug-tight condition prior to pretensioning.
		■	Witness pretensioned and slip critical fastener installations and confirm conformance.
		■	Verify surface condition of faying surfaces attached with pretensioned fasteners.

Source: Ingenium Design

exhaustively detailed in the codes. Additionally, Chapter 10 in *Special Structural Topics*[8] in this series provides an in-depth look at steel quality control.

2.8.1 Field Observations

We frequently visit the construction site to observe progress and handle challenges that arise. An engineer or architect should not direct the work. Rather, according to the contract, he or she should communicate their finding in writing to the contractor, architect, owner, and jurisdiction. Things to look for while in the field include:

- Member size and orientation. If the erection and shop drawings are correct, it is likely this won't be a problem. But it's always worth looking.
- Field holes. Confirm torch burning of holes has not taken place.
- Welding. Observe the welds and see if they are smooth and have a fluid V pattern, as shown in Figure 2.15. If they are bumpy and **discontinuous**, there is a problem.

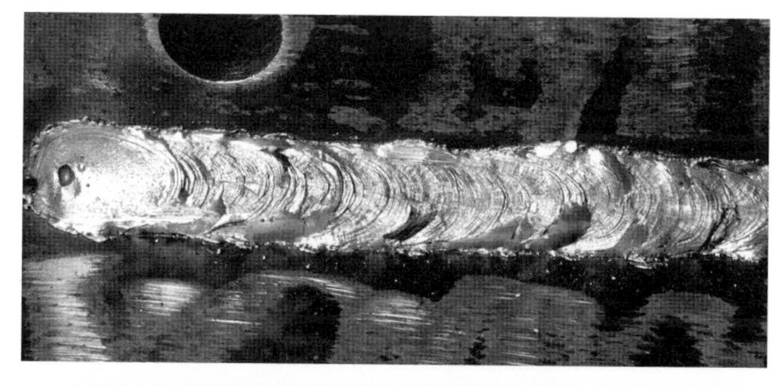

Figure 2.15 (a) Good quality, and (b) poor quality single-pass **fillet**

- Bolts. Confirm all holes are filled with bolts. Observe the bolt head markings and confirm the right grade is used. Spot check the bolt sizes and make sure they follow the plans.
- Bolt Pretensioning. For **pretensioned** and slip critical joints confirm the bolts have been properly tensioned. This might be through direct tension indicator washers or field inspection reports.
- Base plate grouting. Confirm base plates are grouted and it stops at the bottom of the baseplate. If it goes up the sides, it will crack.

2.9 WHERE WE GO FROM HERE

From here we will dive into tension, bending, shear and compression member design. We then get into lateral design, and end with connections. By the time you are finished studying this book, you will be familiar with what structural engineers use 80% of the time in steel design.

NOTES

1. Teran Mitchell. "Structural Materials." In *Introduction to Structures*, edited by Paul W. McMullin and Jonathan S. Price. (New York: Routledge, 2016).
2. AISC, *Specification for Structural Steel Buildings*, AISC 360 (Chicago: American Institute of Steel Construction, 2016).
3. AISC, *Code of Standard Practice for Steel Buildings and Bridges*, AISC 303 (Chicago: American Institute of Steel Construction, 2016).
4. RSSC, *Specification for Structural Joints Using High-Strength Joints* (Chicago: Research Council on Structural Connections, 2014).
5. AISC, *Seismic Provisions for Structural Steel Buildings*, AISC 314 (Chicago: Research Council on Structural Connections, 2016).
6. AISC, *Steel Construction Manual*, 14th edition (Chicago: American Institute of Steel Construction, 2011).
7. AISC, *Steel Construction Manual*.
8. William Komlos. "Quality and Inspection." In *Special Structural Topics*, edited by Paul W. McMullin and Jonathan S. Price. (New York: Routledge, 2017).

Steel Tension

Chapter 3

Richard T. Seelos

You are likely familiar with tension forces. In our common experience, they are present in ropes, cables, and chains. They act to stretch a member along its length, illustrated in Figure 3.1 and Figure 3.2. In structures, we find tension forces in bracing (Figure 3.3), hangers, trusses (Figure 3.4), and beams and columns of lateral systems.

When we design tension members, it is important to follow the load as it transfers from connection to member and back to connection. This is known as **load path**, and can be visualized as a chain, each link a possible source of strength or weakness. Imagine that we pull on the chain shown in Figure 3.5. As we increase the tensile force, we expect the chain to break at some point. As the adage says, the chain is only as strong as the weakest link.

When designing a member, start with the load at one end and track all the elements and members it passes through. Then think of every way they may fail. Each of these failure modes may be considered as a link in

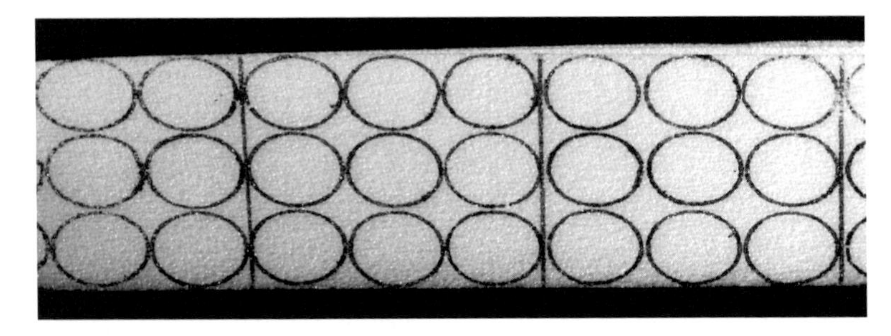

Figure 3.1 Foam tension member showing deformations

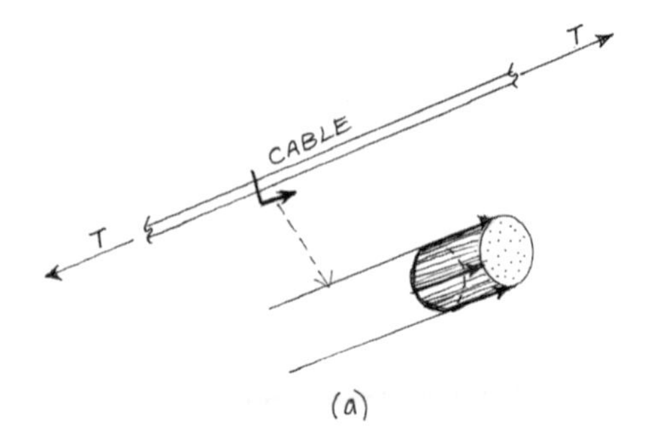

Figure 3.2 Cable under tension load showing uniform stress and deflection

Figure 3.3 Diagonal bracing carrying tension and compression, One Maritime Plaza, San Francisco, SOM

Figure 3.4 Truss carrying tension in bottom chord and diagonals, Harry S. Truman Bridge, Kansas City, Missouri

the chain. For a bolted connection, the failure modes are yield on gross, rupture on net, bolt **bearing**, tear-out, and tear-through, and finally failure of the bolt itself, illustrated in Figure 3.5. Find the weakest link and you have found the tension capacity of the system.

Many elements that make up a building undergo axial tension under different types of loading, shown in Figure 3.6. Some of these elements may be due to **gravity loads**, like a hanging column or truss chord.

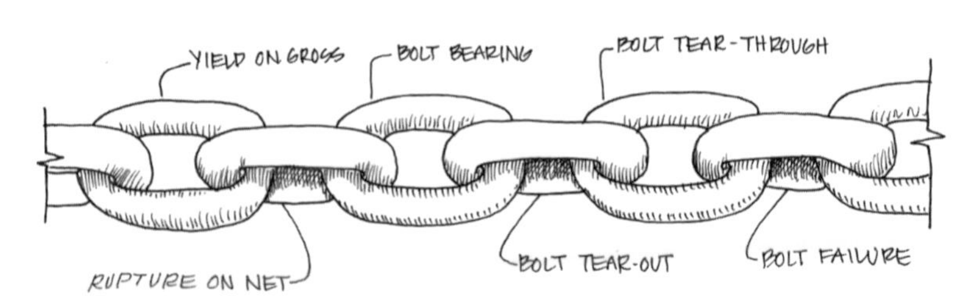

Figure 3.5 Tension as the proverbial chain

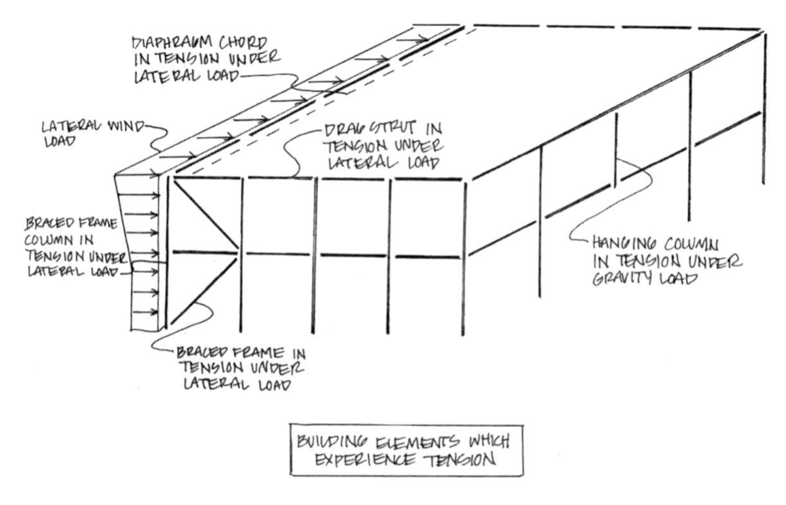

Figure 3.6 Building elements that undergo tension

However, many tension elements occur due to **lateral loading**, which cause reversing tension and compression as the load changes direction. See Chapter 7 for a discussion on lateral load paths and the forces in diaphragms and frames.

3.1 STABILITY

The strength of a tension element is not a function of stability. We never worry about a tension element buckling. Thus, unlike in compression, the element slenderness is not a strength-related issue. However, AISC recommends the following slenderness ratio to prevent a tension member from becoming overly flexible (though this limit does not apply to rods).

$$\frac{L}{r} \leq 300 \tag{3.1}$$

where

L = distance between bracing locations (in, mm)
r = least radius of gyration (in, mm).

3.2 CAPACITY

Recalling our discussion about load paths, the following sections discuss the different tension failure modes. Each mode may be considered a link in

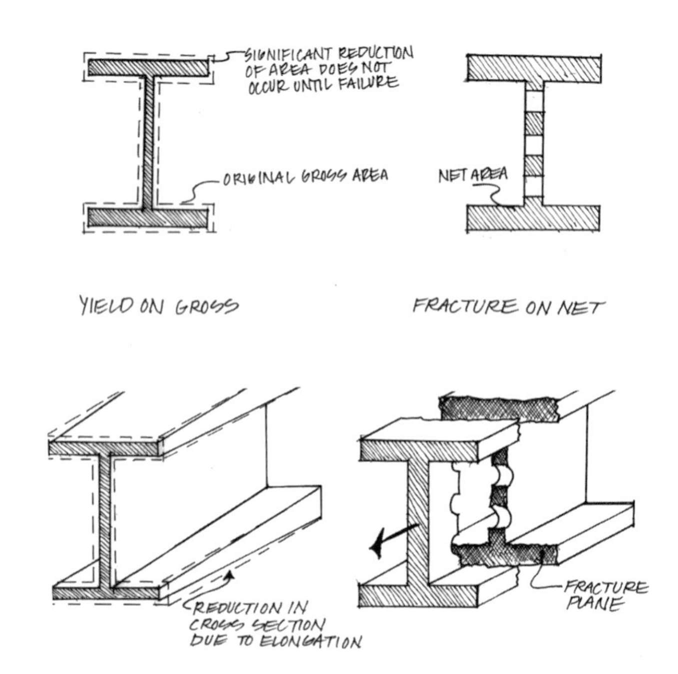

Figure 3.7 Yield on gross and rupture on net

our proverbial chain, and must be checked to determine the capacity of our system. We will focus on failure of the member itself in this chapter, specifically yield and rupture, illustrated in Figure 3.7. We also introduce connections, but cover these extensively in Chapter 8.

3.2.1 Yield on Gross

Yielding on the gross cross-sectional area occurs when the force in the member is large enough to cause the entire area to reach the material yield point. This results in large member elongation with a small increase in force, shown in Figure 2.4.

To determine the nominal yield on gross capacity of an element, we use the following equation.

$$P_n = F_y A_g \tag{3.2}$$

where

F_y = yield stress of material (k/in^2, N/mm^2)

A_g = gross area, in2 in^2 (mm^2)
ϕ = 0.9, strength reduction factor.

Multiplying P_n by ϕ, we get the design strength ϕP_n. We compare this to the force in the member T_u, discussed further below.

3.2.2 Rupture on Net

For members with bolt holes, the gross area is reduced. While the gross area may have stresses less than yield, the stresses at the **net area** may exceed the ultimate (or rupture) strength. (Refer again to the stress-strain diagram shown in Figure 2.4.) We can allow the net area to yield, but we do not want it to rupture. The net area occurs over such a small length of the member that when it yields, the resulting elongation is relatively small.

To determine the nominal rupture on net capacity, we use the following equation.

$$P_n = F_u A_e \tag{3.3}$$

where

F_u = ultimate stress of material, k/in^2 (MN/m^2)
A_e = **effective area**, in^2 (mm^2)
ϕ = 0.75.

It is of great importance that this equation uses the effective area rather than the net area, discussed below.

3.2.2.1 Net Area

Net area accounts for the loss of cross-sectional area due to holes in the member. For welded connections, the net area is typically equal to the gross area. (An exception to this would be plug or slot welds.) For bolted connections, the net area is equal to the gross area minus the area removed by the bolt holes.

There are four common types of bolt holes. Other sizes may be used, but must be dimensioned by the designer. The four types are; standard and oversized holes, and short and long slots. Hole and slot sizes are provided in Chapter 8.

Standard holes are used everywhere except where the designer specifies otherwise. The standard hole is 1/16 in (2–3 mm) larger than the bolt diameter. Oversized holes can be used with slip-critical bolts, but not for bearing connections. Oversized holes vary from 1/8 to 5/16 in

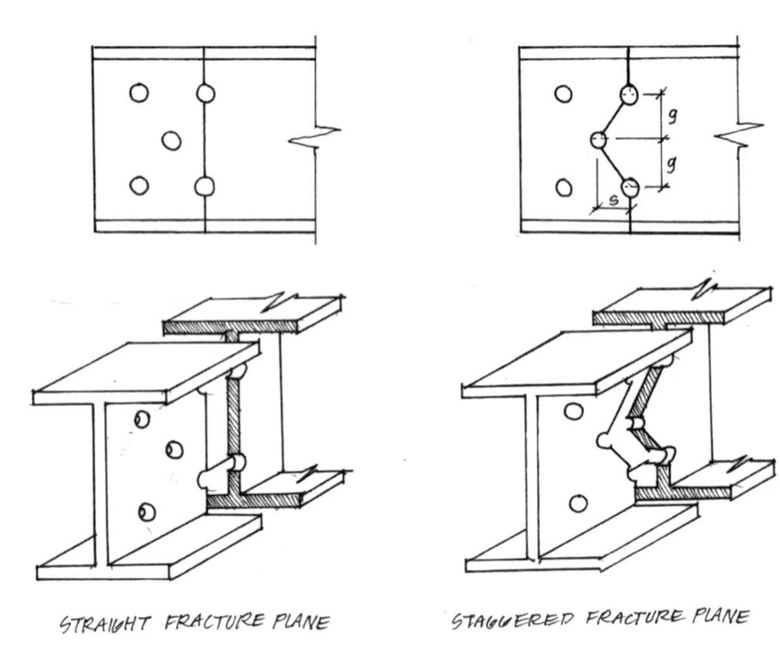

Figure 3.8 Staggered hole failure mode

(4–8 mm) larger than the bolt. Short and long-slotted holes are the same as a standard hole, but elongated in one direction. They can be used for bearing and slip critical connections, but the slot must be perpendicular to the force in bearing connections.

When calculating net area, we take the size of the bolt hole plus 1/16 in (2 mm) to determine the area that needs to be subtracted from the gross area. The additional 1/16 in accounts for drilling imperfections.

When bolt holes are staggered, the critical failure plane may not be a straight line, as illustrated in Figure 3.8. This is further discussed in section B2 of *AISC-360*.[1]

3.2.2.2 Effective Area

Effective area accounts for how a tensile force is distributed from the full cross-section of the member to the area where the connection actually occurs. Also known as **shear lag**, it accounts for force flowing from one part of a member to another, shown in Figure 3.9.

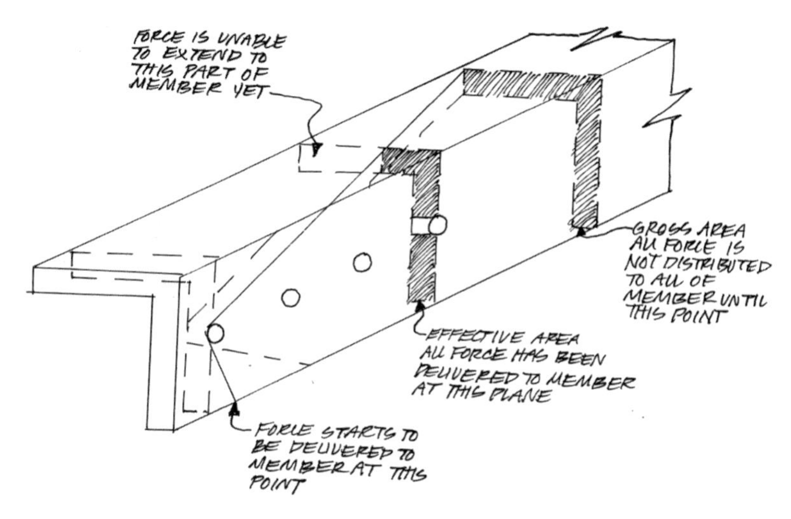

Figure 3.9 Force transfer utilizing shear lag

Effective area is related to net area through the shear lag factor U, as follows:

$$A_e = A_n U \tag{3.4}$$

where

A_n = net area, in² (mm²)
U = shear lag factor (unitless).

The shear lag factor is 1.0, when each portion of a cross-section is connected, like that shown in Figure 3.10. It is less than one when only some portions of a member are connected, like Figure 3.11. Calculation of the shear lag factor can be effort intensive. Table 3.1 provides them for a number of conditions.

Rupture controls for the conditions, for the given strengths

- A992 when $A_e < 0.923\ A_g$
- A36 when $A_e < 0.745\ A_g$
- A500 Grade B when $A_e < 0.952\ A_g$.

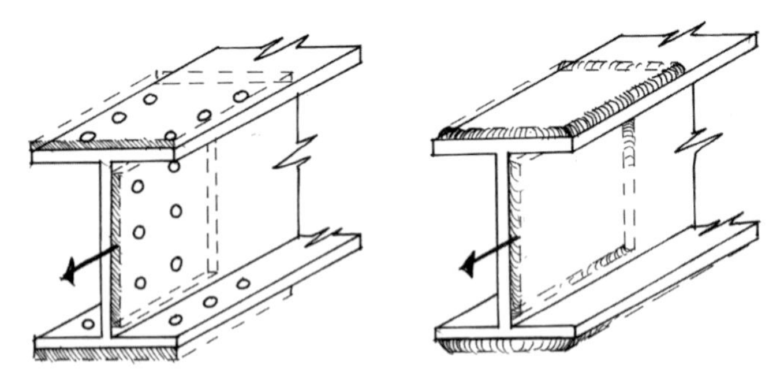

Figure 3.10 Tension transmitted directly to all cross-sectional elements

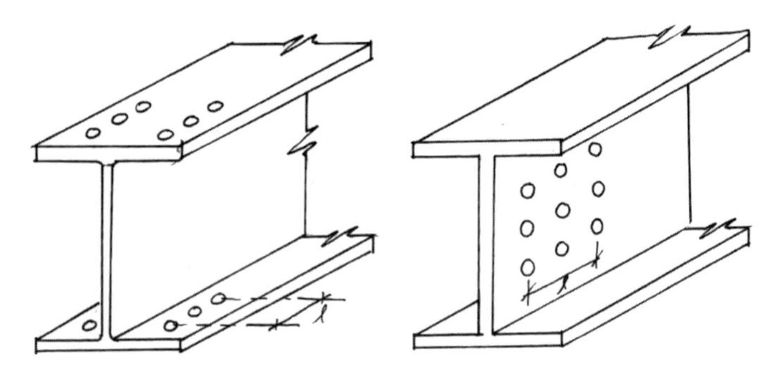

Figure 3.11 Tension transmitted to only some of cross-sectional elements with bolts

3.2.3 Connection Failure Modes

Members must be connected to each other to transfer tension forces, otherwise our proverbial chain would end, as would our load path. Connection failure modes that must be considered when looking at tension load paths include welds, bolts, plates, angles, and connected portions of connected members. These failure modes are extensively covered in Chapter 8, and illustrated in the example in this chapter.

3.2.3.1 Initial Tension Sizing

An effective rule of thumb for initial member sizing is to take the demand and work backwards to area. This will give us an idea of the required section size. Using the following equation, and previously defined terms above, we find the required area as:

Table 3.1 Shear lag factors for various connection types

Case	*Description*	*Shear Lag Factor*	*Sketch*
1	Force is transferred to each portion of a member cross section	$U = 1.0$	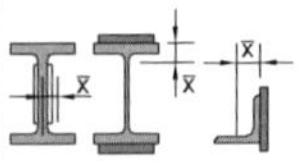
2	Force is transferred to only some of the cross section	$U = 1 - (\bar{x}/l)$	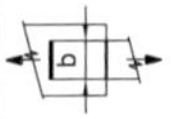
3	Force transmitted by weld perpendicular to force, and area based on connected part *(bt)*	$U = 1.0$	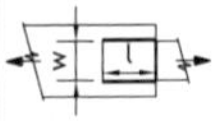
4	Plate where force is transmitted by welds parallel to force	$l \geq 2w,\ U = 1.0$ $2w > l \geq 1.5w,\ U = 0.87$ $1.5w > l \geq 1.5w,\ U = 0.87$	
5	Round HSS with middle gusset	$l \geq 1.3D,\ U = 1.0$ $D \leq l < 1.3D,\ U = 1 - (\bar{x}/l)$ $\bar{x} = D/\pi$	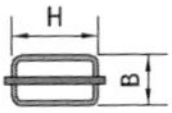
6	Rectangular HSS with middle gusset	$l \geq H,\ U = 1 - (\bar{x}/l)$ $\bar{x} = B^2 + 2BH/4(B + H)$	
7a	W, M, S, HP, or WT with 3 or more rows of fasteners in flanges	U = 0.85	
7b	W, M, S, HP, or WT with 4 or more rows of fasteners in web	U = 0.7	
8a	Angles with 4 or more fasteners parallel to load	U = 0.8	
8b	Angles with 3 fasteners parallel to load	U = 0.6	

Source: AISC 360–16

$$A_{req} = \frac{T_u}{0.9F_y}$$

(3.5)

Another method is to use the tension yield strengths provided in Table 3.2. These can be compared directly with tension force to give an idea of required area. Regardless of approach, we will need to check rupture on net once we select our member and connections.

3.3 DEMAND VS CAPACITY

The LRFD equation for demand versus capacity uses factored load combinations, which produce demand (load) T_u, and compares this with the nominal resistance multiplied by the strength reduction factor ϕ, yielding capacity ϕP_n. When the capacity is greater than the demand, our design works, as shown in Equation 3.6.

$$\phi P_n \geq T_u$$

(3.6)

The strength reduction factor ϕ is 0.9 for yielding and 0.75 for limit states involving rupture, like rupture on net.

Looking at demand more in-depth, we see demand T_u is defined as in Equation 3.7:

$$T_u = \sum \gamma_i Q_i$$

(3.7)

where

γ_I = Load Factor
Q_i = Nominal Load, k (kN).

Load factors, which are a part of load combinations, are discussed further in *Introduction to Structures* in this series, and *ASCE7*.[2]

3.4 DEFLECTION

There are two main classes of limit states, strength and serviceability. We use **ASD** load combinations when checking serviceability. Deflection is the main serviceability limit state we are concerned with when looking at tension element design.

The code does not have any specific tension deflection requirements, but they can be determined by considering the types of systems they are in and any adverse effects that would result from excessive deformation. For example, if a brace in a vertical frame were to elongate excessively,

Table 3.2 Yield on gross capacities based on area

Gross Area (in^2)	*Tension Strength* ϕP_n *(k)* *Yield Strength* f_y (k/in^2)		
	36	*46*	*50*
0.500	16.2	20.7	22.5
0.800	24.3	31.1	33.8
1.00	32.4	41.4	45.0
1.50	48.6	62.1	67.5
2.00	64.8	82.8	90.0
3.00	97.2	124	135
4.00	130	166	180
5.00	162	207	225
7.50	243	311	338
10	324	414	450
12.5	405	518	563
15.0	486	621	675
20.0	648	828	900
30.0	972	1,242	1,350
40.0	1,296	1,656	1,800
50.0	1,620	2,070	2,250
70.0	2,268	2,898	3,150
100	3,240	4,140	4,500
125	4,050	5,175	5,625
150	4,860	6,210	6,750
175	5,670	7,245	7,875
200	6,480	8,280	9,000
250	8,100	10,350	11,250
300	9,720	12,420	13,500
400	12,960	16,560	18,000
500	16,200	20,700	22,500

Table 3.2m Yield on gross capacities based on area

Gross Area (mm^2)	*Tension Strength ϕP_n (kN)* *Yield Strength fy (MN/m^2)*		
	250	*317*	*345*
320	72.0	91.3	99.4
480	108	137	149
650	146	185	202
970	218	277	301
1,300	293	371	404
2,000	450	571	621
2,580	581	736	801
3,200	720	913	994
4,800	1,080	1,369	1,490
6,500	1,463	1,855	2,018
8,100	1,823	2,311	2,515
9,700	2,183	2,767	3,012
13,000	2,925	3,709	4,037
19,400	4,365	5,535	6,024
25,800	5,805	7,361	8,011
32,300	7,268	9,215	10,029
45,200	10,170	12,896	14,035
64,500	14,513	18,402	20,027
80,700	18,158	23,024	25,057
96,800	21,780	27,617	30,057
113,000	25,425	32,239	35,087
129,000	29,025	36,804	40,055
161,000	36,225	45,933	49,991
194,000	43,650	55,348	60,237
258,000	58,050	73,607	80,109
323,000	72,675	92,152	100,292

Note 1) Check net section rupture for members with holes

the resulting story **drift** may be unacceptable. Similarly, the horizontal deflection limits discussed in Chapter 4 for beams, can be applied to truss deflections, which are the result of tension and compression deflections of the chords and webs.

From strengths and materials, we know axial deflection δ is given as in Equation 3.8.

$$\delta = \frac{PL}{A_g E} \tag{3.8}$$

where

P = force, k (kN)
L = length of member, in (mm)
A_g = gross area of the element, in^2 (mm^2)
E = modulus of elasticity, k/in^2 (MN/mm^2).

Note this equation only holds true in the elastic range. As discussed in Chapter 2, once a stress exceeds the yield strength the elongation is no longer linearly related to deflection.

3.5 DETAILING

When **detailing** tension elements, keep in mind the effective area of the connection, which is a function of net area and shear lag. We balance bolt size with member size, so we have about a 70% reduction in area. For shear lag, it is best to connect each portion of the member—both legs of an angle or both flanges and web of wide flanges. When this is not feasible or economical, make the length of the connection sufficiently long to avoid large reductions in effective area, and thus connection capacity. Connection detailing is further discussed in Chapter 8.

3.6 DESIGN STEPS

To design a tension member, we must determine the following information:

Step 1: Determine the structural layout
Step 2: Determine the loads
  a. Maximum **service load**
  b. Maximum factored load
Step 3: Determine material parameters
Step 4: Determine initial size

3.6 DESIGN STEPS *(continued)*

Step 5: Check All Applicable Strength Limit States
- a. Yield on Gross
- b. Rupture on Net
- c. Bolt Bearing
- d. Bolt Tear-out
- e. **Block Shear**

Step 6: Check Member Serviceability Limit States
- a. Check slenderness requirements
- b. Check deflection requirements

Step 7: Summarize the Final Results

3.7 EXAMPLE: TENSION MEMBER DESIGN

Step 1: Determine the Structural Layout

Our structural configuration is shown in Figure 3.12. We will design the steel beam for tension, which collects lateral seismic forces and drags it into an adjacent steel **braced frame**. Known as a drag strut, it is 28 ft (8.5 m) long, and connected with plates at the top and bottom flanges, with 3/4 in (20 mm) diameter bolts in standard holes, illustrated in Figure 3.12.

We will limit this example to the tension design of the drag strut. Perhaps once you have finished the book, you can check the bending and compression capacity of the beam, and failure modes of the bolts, drag plates, welds. Thus we will only be looking at one section of the chain, and its associated links. In reality, the chain extends from the surface of the element where the load is first applied all the way to the **footings** where it is resolved by the soil beneath the building. It simplifies our design to break the load path into smaller discrete sections that can be evaluated.

Step 2: Determine the Loads

Axial loads in our drag strut are only created by seismic and wind. From our **structural analysis**, we determine that the seismic and wind axial loads are shown in Equation 3.9.

$T_E = 200$ k	$T_E = 0.89$ MN
$T_w = 240$ k	$T_w = 1.064$ MN

Determine the maximum factored load.

We will check the seismic and wind combinations, since this is where our load is coming from. For seismic:

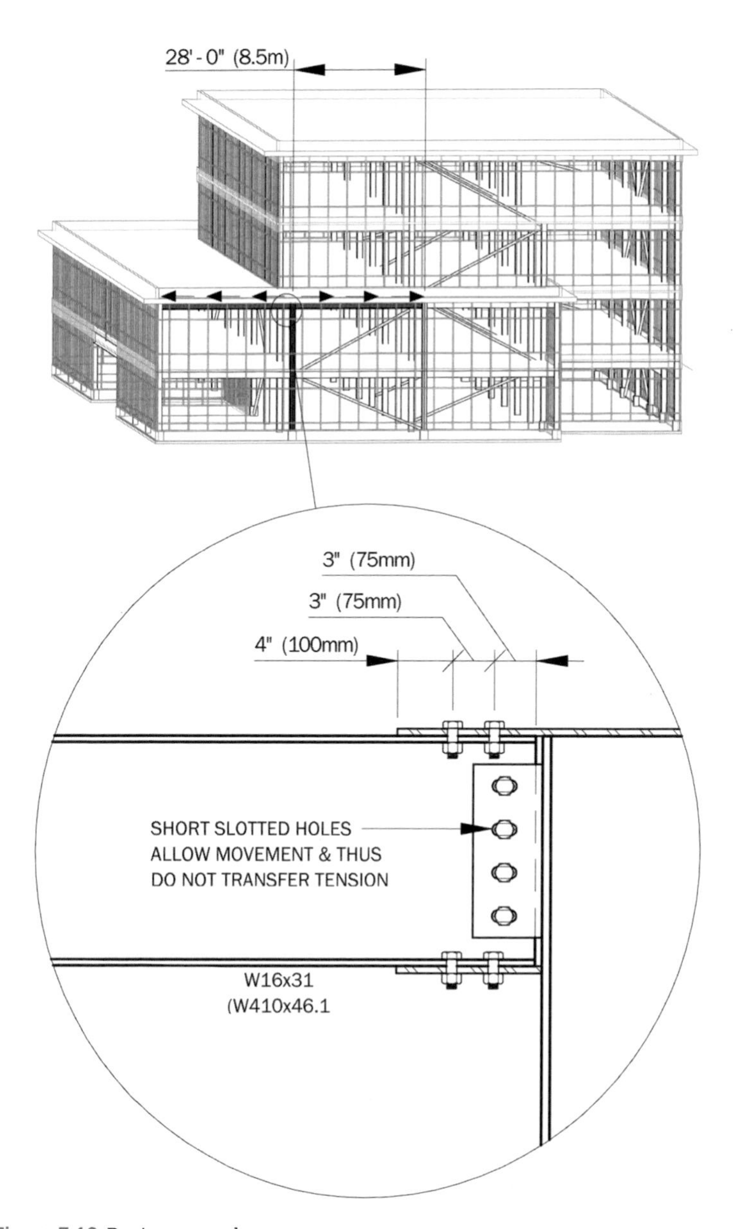

Figure 3.12 Design example

$T_u = 0.9D \pm 1.0E$	
$= 0 + 1.0\ (200.0\ k)$	$= 0 + 1.0\ (0.890\ MN)$
$= 200.0\ k$	$= 0.890\ MN$

For wind we have

$T_u = 0.9D \pm 1.0W$	
$= 0 + 1.0\ (240.0\ k)$	$= 0 + 1.0\ (1.064\ MN)$
$= 240.0\ k$	$= 1.064\ MN$

It looks like wind controls.

Step 3: Determine Material Parameters

The typical material for a wide flange beam is A992. It has a yield strength $F_y = 50\ k/in^2$ ($345\ N/mm^2$) and an ultimate strength $F_u = 65\ k/in^2$ ($450\ N/mm^2$).

Step 4: Determine Initial Size

Because this member will also take bending forces, we estimate its depth as half the beam span in feet, giving us 14 in (356 mm).

Additionally, the tension attachments are plates bolting to the top flange of the beam. The beam flange will need to be wide enough for proper bolt clearances. The required clearance is 1.25 in (31.8 mm), from Table 8.11. The nut diameter is 1.5 in (38.1 mm) and we can approximate the rounding at the corner between the flange and the web as 0.75 in (19.1 mm). Thus, the minimum flange width is:

2(1.25 in+1.5 in /2+0.75 in)	2(31.8 mm+38.1 mm/2+19.1 mm)
= 5.5 in	= 140 mm

Based on these criteria, we choose a W16 × 31 (W410 × 46.1), which has a beam width of 5.53 in (140 mm).

Step 5: Check Strength Limit States

Step 5a: Yield on Gross

From the steel beam properties table for a W16 × 31, the area is.

$A_g = 9.13\ in^2$	$A_g = 5.890\ mm^2$

The tensile yield strength follows as

$\phi P_n = \phi F_y A_g$	
$= 0.9(50\ k/in^2)(9.13\ in^2)$	$= 0.9(345\ N/mm^2)(5{,}890\ mm^2)\left(\dfrac{1\ MN}{1\times10^6\ N}\right)$
$= 411\ k$	$= 1.83\ MN$

This is quite a bit larger than our demand. Let's check rupture.

Step 5B: Rupture on Net

We begin by finding the net area, given as:

$$A_n = A_g - A_{\text{hole}}$$

$$= 9.13\ \text{in}^2 - 4(0.44\ \text{in})\left(\frac{13}{16}\ \text{in} + \frac{1}{16}\ \text{in}\right) \qquad = 5{,}890\ \text{mm}^2 - 4(11.2\ \text{mm})(22\ \text{mm} + 2\ \text{mm})$$

$$= 7.59\ \text{in}^2 \qquad = 4{,}815\ \text{mm}^2$$

Because only the flanges are connected, we need to determine the shear lag factor U. From Table 3.1, line 7a, we get $U = 0.85$ for three bolts in a row. The effective area then becomes

$$A_e = A_n U$$

$$= 7.59\ \text{in}^2(0.85) \qquad = 4{,}815\ \text{mm}^2(0.85)$$

$$= 6.45\ \text{in}^2 \qquad = 4{,}093\ \text{mm}^2$$

Calculating the axial tension rupture capacity we get

$$\phi P_n = \phi_t F_u A_e$$

$$= 0.75(65\ \text{k/in}^2)(6.45\ \text{in}^2) \qquad = 0.75(450\ \text{N/mm}^2)(4{,}093\ \text{mm}^2)\left(\frac{1\ \text{MN}}{1 \times 10^6\ \text{N}}\right)$$

$$= 314\ \text{k} \qquad = 1.38\ \text{MN}$$

Step 5C: Compare Strength Limit States

Because both the yield and rupture strengths ϕP_n are greater than the demand T_u, we know our member is OK.

Step 6: Check Deflection

We will skip the deflection check of this member for now. When we check drift in the structural analysis, it will capture the effect of the drag struts axial extension on the frame system.

Step 7: Summarize the Final Results

The final tension design is a W16 × 31 (W410 × 46.1), A992 steel beam 28 ft (8.5 m) long connected with (12) 3/4 in (20 mm) diameter bolts in standard holes connected to the beam flanges, sketched in Figure 3.12. Remember, we will need to also size this member for combined tension and bending, and compression and bending, in addition to checking the connection.

3.8 WHERE WE GO FROM HERE

From here we will get further into connection failure modes, and their influence on member capacity. Key to this is block shear, where a portion of a connected member fails in a combination of tension and shear, discussed in Chapter 8.

For additional design aids, the AISC *Steel Construction Manual*.[3] includes numerous tables for quick tension design of common shapes.

NOTES

1. AISC, *Steel Construction Manual*, 14th Edition (Chicago: American Institute of Steel Construction, 2011).
2. ASCE, *Minimum Design Loads for Buildings and Other Structures*, ASCE/SEI 7–16 (Reston, VA: American Society of Civil Engineers, 2016).
3. AISC, *Steel Construction Manual*, 14th Edition (Chicago: American Institute of Steel Construction, 2011).

Steel Bending

Chapter 4

Richard T. Seelos

Bending members make up the majority of steel elements in a structure. For every column there are numerous beams and **girders**, like those in Figure 4.1. A key goal of the structural engineer is to reduce the total weight of steel while providing adequate strength and stiffness.

Steel **flexural** elements have three primary failure mechanisms; yielding, **lateral torsional buckling**, and local buckling. Yielding occurs when the beam stress equals the yield stress. It occurs in fully braced members. Lateral torsional buckling is when the beam rolls over in the middle, reducing unbraced member capacity below yield. This is because the steel shapes are optimized to the point where numerous buckling failure modes are possible as the member stress increases. Local buckling occurs in slender shapes, and reduces bending strength.

Before we get into the details, let's visualize how beams deform. Simple span beams with downward loads experience tension in the bottom and compression in the top at the mid-span. Figure 4.2a shows a simply supported foam beam with a **point load** in the middle. Notice how the circles on the bottom middle stretch—indicating tension—and shorten at the top—indicating compression. Taking this further, Figure 4.2b shows a multi-span beam with point loads. The middle deformation is the same as a simply supported beam, but over the middle supports, where the beam is continuous, the tension and compression change places—tension on top, compression on bottom.

4.1 STABILITY

To be efficient, steel beams are deeper than they are wide. In wide flanges and channels, the material is optimized to be as far away from the **neutral**

Figure 4.1 Beams and girders, Salt Lake Public Safety Building, GSBS Architects

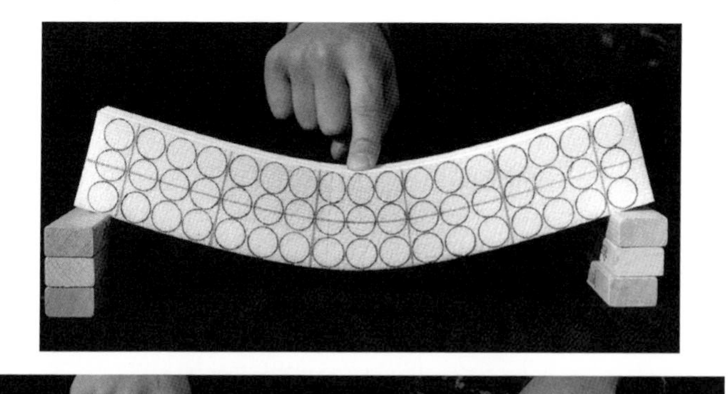
(a)

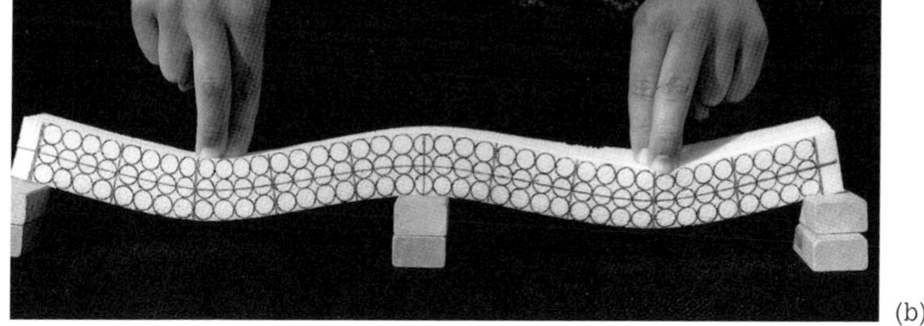
(b)

Figure 4.2 Foam beam showing deformation of (a) single and (b) double span

axis (or center) as possible. This increases the moment of inertia *I*, and plastic section modulus *Z*. (Recall from Table 2.12 the depth is cubed and squared, respectively for these properties, while the **width** is not.) Unfortunately, deeper beams are prone to rolling over in the middle if they are not braced sufficiently.

To understand **lateral torsional buckling**, imagine you begin standing on the edge of a long skinny board, braced at the ends. Applying your weight, it begins to roll over to a flat position near the middle, shown in Figure 4.3. As this happens, the strength of the beam drops rapidly and failure occurs. This is further illustrated in Figure 4.4, which shows a channel in a buckled state under the applied loads.

To keep this from happening, we need to either keep the beam proportions closer to square, or brace it. We generally brace steel beams using one of the following methods:

- Headed studs, like those in Figure 2.14, which provide continuous bracing

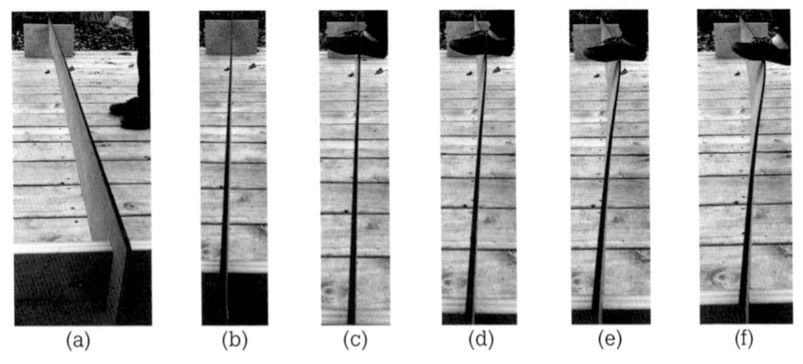

Figure 4.3 Lateral torsional buckling of a slender, unbraced beam showing (a, b) original condition and (c–f) increasing levels of buckling

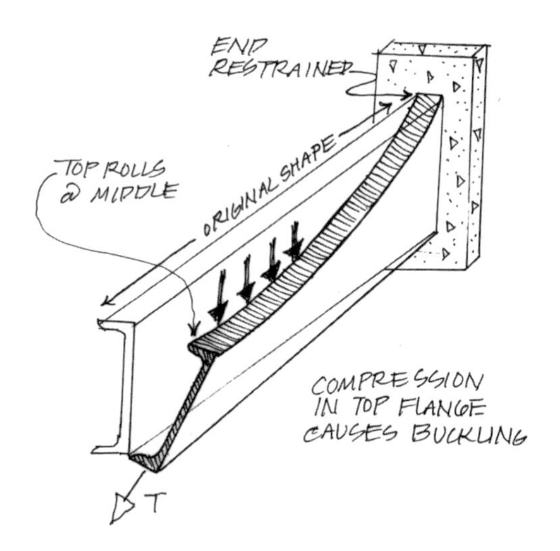

Figure 4.4 Lateral torsional buckling shape of channel

- Joist or Beam framing perpendicular to the beam in question, shown in Figure 4.5a, bracing the compression flange
- Cross-frame bracing, common in highway bridges, shown in Figure 4.5b.

It is important to design the bracing members and their connections for the stability force, which ranges from 2–8% of the bending demand M_u divided by the distance between flanges h_o.

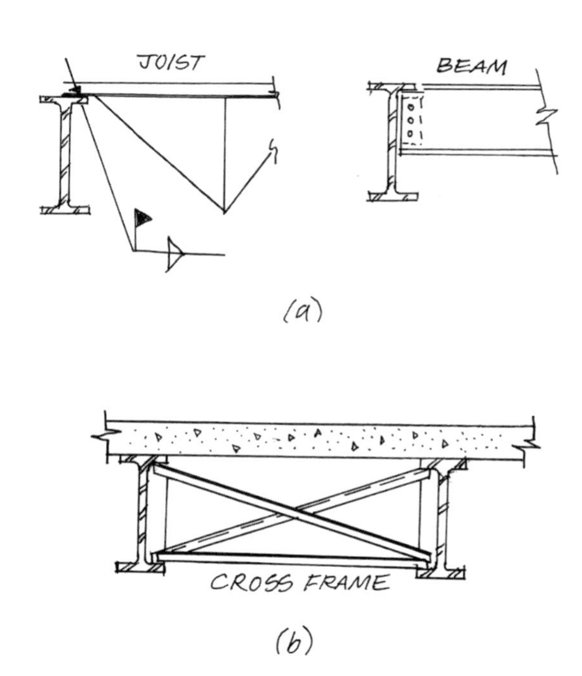

Figure 4.5 Beam bracing by (a) joist or beam and (b) cross frame

4.2 CAPACITY

4.2.1 Bending Stresses

Bending forces in a beam are measured as **moments**, which is a force times a distance to create a rotational torque. For example, if you were turning a wrench and needed more rotational torque, you could apply more force where you are currently pushing or adjust your grip further up the wrench and apply the same amount of force.

These moments are resisted by stresses in the beam. Think of the beam as being made up of a whole bunch of horizontal fibers (similar to a wood beam), each carrying a force, illustrated in Figure 4.6. Because each fiber is located a distance from the centroid of the section, they create a resisting moment. If we sum up all these forces multiplied by the distance at which they act, we can determine the resisting bending moment in the beam.

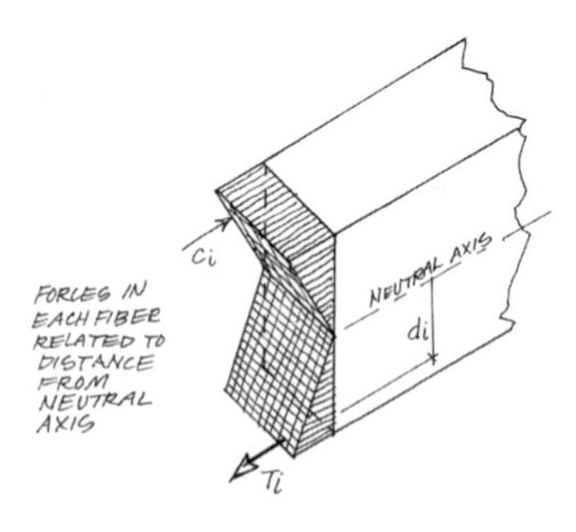

Figure 4.6 Fiber analogy in beam bending

Without moment strength, a beam would not be able to transfer the force from its center to the ends. It would be like trying to use a bicycle chain to support a load between two points as opposed to using a paint stir stick.

The key to summing these incremental forces is to understand the magnitude of stress inside the beam. We do this with strain. Based on the simplified (assumed for calculation) stress strain diagram for steel shown in Figure 2.6, the stress is proportional to the strain as long as the stresses remain in the elastic range—or strain is less than yield strain ε_y. Once the strain exceeds the yield, it remains constant at the yield stress.

With this understanding, let's look at the strain and the stress in a beam as we initiate bending. Starting with a beam that is at the yield strain at the top and bottom, the strain and stress are both linear, shown in Figure 4.7a.

When the strain exceeds the yield strain, the outmost fibers yield and continue to deform, as shown in Figure 4.7b. However, the stress remains constant.

The strain continues to increase until, for all intents and purposes, all the steel yields, like in Figure 4.7c. This is called the full **plastic** capacity of the section. Any additional load will cause the beam to rotate as if there was a hinge at this location. This is called a plastic hinge. The beam now becomes unstable, much like a bicycle chain, and cannot support any additional load.

At this point, all the area above the neutral axis is at yield in compression, while the area below is at yield in tension. Note that because both the sum of the compressive force must equal the tension

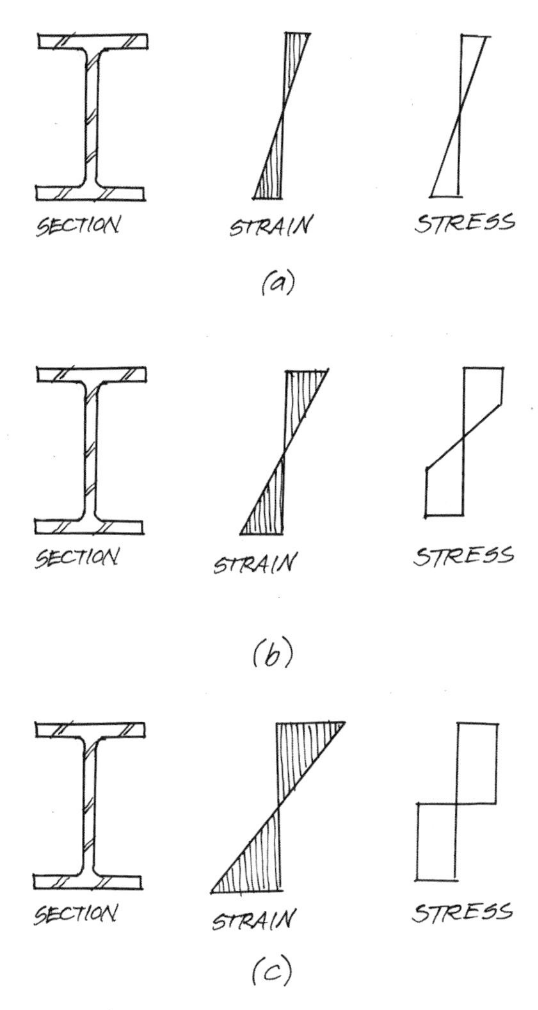

Figure 4.7 Beam stress distribution at (a) yield, (b) past yield, and (c) assumed fully yielded

force and the stress is constant, the area above the plastic neutral axis must equal the area below.

Assuming the yield stress in the beam is constant after yield is a simplification that ignores the stress increase due to strain hardening, as seen on the engineering stress strain diagram in Figure 2.6. However, this provides an additional degree of safety.

To bring this full circle, let's look at the section properties that relate to bending strength. These are the elastic and plastic section moduli S and Z, respectively. They essentially sum up the effect of each individual fiber in the beam, according to the stress states in Figure 4.7a and Figure 4.7c. Multiplying these by yield stress gives us the elastic and plastic bending strength, M_y and M_p, respectively. In equation form this is elastic (Eqn 4.1) and plastic (Eqn 4.2)

$$M_y = SF_y \tag{4.1}$$

$$M_p = ZF_y \tag{4.2}$$

where

S = elastic section modulus, in^3 (mm^3)
Z = plastic section modulus, in^3 (mm^3)
F_y = yield strength of the element, k/in^2 (MN/m^2).

The tables in Appendix 1 provide the elastic and plastic section moduli for a variety of shapes.

4.2.2 Yielding

In beam design, we are concerned with three modes of behavior; yielding, inelastic, and elastic lateral torsional buckling. The modes are based on the distance between bracing and how prone the section is to lateral torsional buckling. The trend is for lower moment strength with larger brace **spacing**, illustrated in Figure 4.8.

The first range of behavior occurs when the beam is sufficiently braced to cause yielding. At this point the nominal bending capacity is equal to the plastic capacity M_p, as in Eqn 4.3.

$$M_n = ZF_y \tag{4.3}$$

Later, we will apply the strength reduction factor ϕ of 0.9.

4.2.3 Lateral Torsional Buckling

For members that are not sufficiently braced, the bending strength begins to decrease due to lateral torsional buckling. This effect is shown in Figure 4.8, which shows two types of buckling; inelastic and elastic lateral torsional buckling (LTB). The transition between yielding and

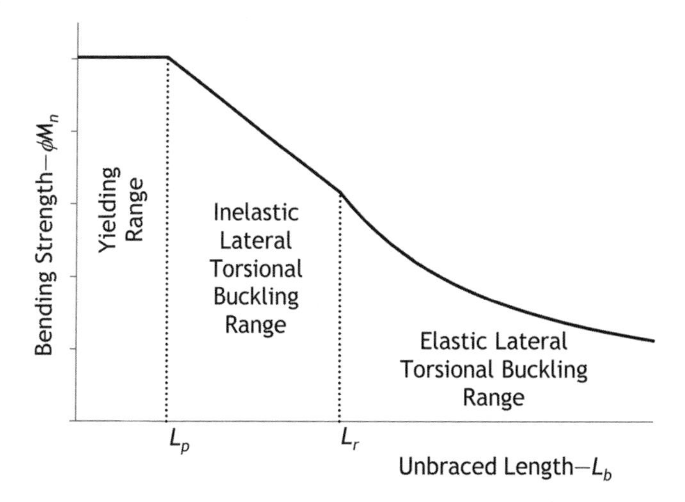

Figure 4.8 Three regions of moment strength, as a function of unbraced length

inelastic LTB occur when braces are spaced at the yielding **unbraced length** L_p. The transition between inelastic and elastic buckling occurs at the inelastic unbraced length L_r. For I-shaped members including hybrid sections and channels, these lengths are defined by equations 4.4 and 4.5:

$$L_p = 1.76 r_y \sqrt{\frac{E}{F_y}} \tag{4.4}$$

$$L_r = 1.95 r_{ts} \frac{E}{0.7F_y} \sqrt{\frac{Jc}{S_x h_o} + \sqrt{\left(\frac{Jc}{S_x h_o}\right)^2 + 6.76\left(\frac{0.7F_y}{E}\right)^2}} \tag{4.5}$$

r_y = **weak axis** radius of gyration, in (mm)
E = modulus of elasticity, k/in^2 (MN/m^2)
F_y = yield strength, k/in^2 (MN/m^2)
r_{ts} = effective radius of gyration, in (mm), given in Eqn 4.6:

$$r_{ts} = \sqrt{\frac{\sqrt{I_y C_w}}{S_x}} \tag{4.6}$$

J = torsional property, in^4 (mm^4)
c = 1 for doubly symmetric I-shapes and

$$C = \frac{h_o}{2}\sqrt{\frac{I_y}{C_w}} \tag{4.7}$$

for channels

I_y = moment of inertia taken about the y **axis**, in^4 (mm^4)
S_x = elastic section modulus taken about the x-axis, in^3 (mm^3)
h_o = distance between the flange centroids, in (mm)
C_w = Warping constant, in^6 (mm^6).

These equations get progressively more involved for other types of shapes, and are found in *AISC 360*. Table 4.1 lists L_p and L_r for selected bending members. We then compare these with the actual brace spacing L_b, to determine in which range of behavior we find our beam.

When $L_p < L_b \leq L_r$ we are in the inelastic LTB range. The beam partially yields and partially buckles laterally. For compact, doubly symmetric shapes and channels, the nominal flexural capacity M_n equals:

$$M_n = C_b\left[M_p - \left(M_p - 0.7F_yS_x\right)\left(\frac{L_b - L_p}{L_r - L_p}\right)\right]$$
$$M_n \leq M_p \tag{4.8}$$

where

C_b = lateral torsional buckling factor, unitless, no less than 1.0
M_p = plastic bending moment, k-ft (kN-m)
L_b = unbraced length, ft (m)
L_p = yielding lateral unbraced length, ft (m)
L_r = inelastic lateral unbraced length, ft (m)
ϕ = strength reduction factor equals 0.9.

Similar equations are available for non-compact sections, $L_b > L_r$. It is recommended to avoid design in the elastic LTB range as no yielding would occur at failure. This would result in an instantaneous failure with no warning to the building inhabitants.

The equation for inelastic LTB capacity shown above is based off the assumption that the beam has a constant moment across the full length of the beam. In truth, this is the exception to the rule, and the moment typically varies over the length of the beam. If the moment is less at various locations along the beam, there will be less force pushing the

Table 4.1 Wide flange bending strength as a function of Z_x

Shape	Z_x (in^3)	ϕM_{px} (k-ft)	ϕM_{rx} (k-ft)	L_p (ft)	L_r (ft)	I_x (in^4)
W36 × 652	2,910	10,900	6,460	14.5	77.7	50,600
W40 × 593	2,760	10,400	6,140	13.4	63.9	50,400
W27 × 539	1,890	7,090	4,120	12.9	88.5	25,600
W40 × 392	1,710	6,410	3,780	9.33	38.3	29,900
W14 × 730	1,660	6,230	3,360	16.6	275	14,300
W44 × 335	1,620	6,080	3,700	12.3	38.9	31,100
W33 × 387	1,560	5,850	3,540	13.3	53.3	24,300
W30 × 391	1,450	5,440	3,280	13	58.8	20,700
W24 × 370	1,130	4,240	2,510	11.6	69.2	13,400
W36 × 210	833	3,120	1,890	9.11	28.5	13,200
W33 × 201	773	2,900	1,800	12.6	36.7	11,600
W18 × 311	754	2,830	1,640	10.4	81.1	6,970
W30 × 211	751	2,820	1,750	12.3	38.7	10,300
W27 × 217	711	2,670	1,650	11.7	40.8	8,910
W12 × 336	603	2,260	1,270	12.3	150	4,060
W40 × 149	598	2,240	1,350	8.09	23.6	9,800
W21 × 201	530	1,990	1,210	10.7	46.2	5,310
W36 × 135	509	1,910	1,150	8.41	24.3	7,800
W14 × 257	487	1,830	1,090	14.6	104	3,400
W24 × 162	468	1,760	1,090	10.8	35.8	5,170
W33 × 118	415	1,560	942	8.19	23.4	5,900
W12 × 230	386	1,450	843	11.7	105	2,420
W18 × 158	356	1,340	814	9.68	42.8	3,060
W21 × 122	307	1,150	717	10.3	32.7	2,960
W30 × 90	283	1,060	643	7.38	20.9	3,610
W24 × 103	280	1,050	643	7.03	21.9	3,000

Table 4.1 *continued*

Shape	Z_x (in^3)	ϕM_{px} (k-ft)	ϕM_{rx} (k-ft)	L_p (ft)	L_r (ft)	I_x (in^4)
W27 × 84	244	915	559	7.31	20.8	2,850
W12 × 152	243	911	549	11.3	70.6	1,430
W16 × 100	198	743	459	8.87	32.8	1,490
W24 × 68	177	664	404	6.61	18.9	1,830
W16 × 89	175	656	407	8.8	30.2	1,300
W14 × 99	173	646	412	13.5	45.3	1,110
W21 × 73	172	645	396	6.39	19.2	1,600
W12 × 106	164	615	381	11.0	50.7	933
W18 × 76	163	611	383	9.22	27.1	1,330
W14 × 90	157	574	375	15.1	42.5	999
W12 × 96	147	551	344	10.9	46.7	833
W10 × 112	147	551	331	9.47	64.1	716
W18 × 71	146	548	333	6.00	19.6	1,170
W24 × 55	134	503	299	4.73	13.9	1,350
W21 × 50	110	413	248	4.59	13.6	984
W16 × 57	105	394	242	5.65	18.3	758
W18 × 50	101	379	233	5.83	16.9	800
W12 × 65	96.8	356	231	11.9	35.1	533
W21 × 44	95.4	358	214	4.45	13.0	843
W12 × 58	86.4	324	205	8.87	29.8	475
W12 × 53	77.9	292	185	8.76	28.2	425
W12 × 50	71.9	270	169	6.92	23.8	391
W8 × 67	70.1	263	159	7.49	47.6	272
W14 × 43	69.6	261	164	6.68	20.0	428
W18 × 35	66.5	249	151	4.31	12.3	510
W16 × 36	64.0	240	148	5.37	15.2	448

Table 4.1 *continued*

Shape	Z_x *(in³)*	ϕM_{px} *(k-ft)*	ϕM_{rx} *(k-ft)*	L_p *(ft)*	L_r *(ft)*	I_x *(in⁴)*
W10 × 49	60.4	227	143	8.97	31.6	272
W12 × 40	57.0	214	135	6.85	21.1	307
W12 × 35	51.2	192	120	5.44	16.6	285
W16 × 26	44.2	166	101	3.96	11.2	301
W10 × 33	38.8	146	91.9	6.85	21.8	171
W12 × 26	37.2	140	87.7	5.33	14.9	204
W14 × 22	33.2	125	76.1	3.67	10.4	199
W10 × 22	26.0	97.5	60.9	4.7	13.8	118
W12 × 19	24.7	92.6	55.9	2.9	8.61	130
W8 × 24	23.1	86.6	54.9	5.69	18.9	82.7
W12 × 14	17.4	65.3	39.1	2.66	7.73	88.6
W8 × 18	17.0	63.8	39.9	4.34	13.5	61.9
W10 × 12	12.6	46.9	28.6	2.87	8.05	53.8
W8 × 10	8.87	32.9	20.5	3.14	8.52	30.8

Source: AISC Steel Construction Manual, 14th Edition

beam into the lateral torsional buckling mode. We capture this with the bending coefficient C_b, given as

$$C_b = \frac{12.5M_{\max}}{2.5M_{\max}+3M_A+4M_B+3M_C} \tag{4.9}$$

M_{max} = absolute value of maximum moment in unbraced segment (kip-ft)

M_A = absolute value of moment at quarter point of unbraced segment (kip-ft)

M_B = absolute value of moment at centerline of unbraced segment (kip-ft)

M_C = absolute value of moment at three-quarter point of unbraced segment (kip-ft).

Table 4.1m Wide flange bending strength as a function of Z_x

Shape	Z_x / 10^3 (mm^3)	ϕM_{px} (kN-m)	ϕM_{rx} (kN-m)	L_p (m)	L_r (m)	I_x / 10^6 (mm^4)
W920 × 970	47,686	14,778	8,758	4.42	23.7	21,100
W1000 × 883	45,228	14,100	8,325	4.08	19.5	21,000
W690 × 802	30,972	9,613	5,586	3.93	27.0	10,700
W1000 × 584	28,022	8,691	5,125	2.84	11.7	12,400
W360 × 1086	27,203	8,447	4,555	5.06	83.8	5,950
W1100 × 499	26,547	8,243	5,016	3.75	11.9	12,900
W840 × 576	25,564	7,931	4,800	4.05	16.2	10,100
W760 × 582	23,761	7,376	4,447	3.96	17.9	8,620
W610 × 551	18,517	5,749	3,403	3.54	21.1	5,580
W920 × 313	13,650	4,230	2,562	2.78	8.69	5,490
W840 × 299	12,667	3,932	2,440	3.84	11.2	4,830
W460 × 464	12,356	3,837	2,224	3.17	24.7	2,900
W760 × 314	12,307	3,823	2,373	3.75	11.8	4,290
W690 × 323	11,651	3,620	2,237	3.57	12.4	3,710
W310 × 500	9,881	3,064	1,722	3.75	45.7	1,690
W1000 × 222	9,799	3,037	1,830	2.47	7.29	4,080
W530 × 300	8,685	2,698	1,641	3.26	14.1	2,210
W920 × 201	8,341	2,590	1,559	2.56	7.41	3,250
W360 × 382	7,981	2,481	1,477	4.45	31.7	1,420
W610 × 241	7,669	2,386	1,478	3.29	10.9	2,150
W840 × 176	6,801	2,115	1,277	2.50	7.13	2,460
W310 × 342	6,325	1,966	1,143	3.57	32.0	1,010
W460 × 235	5,834	1,817	1,104	2.95	13.0	1,270
W530 × 182	5,031	1,559	972	3.14	9.97	1,230
W760 × 134	4,638	1,437	872	2.25	6.37	1,500
W610 × 153	4,588	1,424	872	2.14	6.68	1,250

Table 4.1m *continued*

Shape	Z_x / 10^3 (mm^3)	ϕM_{px} (kN-m)	ϕM_{rx} (kN-m)	L_p (m)	L_r (m)	I_x / 10^6 (mm^4)
W690 × 125	3,998	1,241	758	2.23	6.34	1,190
W310 × 226	3,982	1,235	744	3.44	21.5	595
W410 × 149	3,245	1,007	622	2.70	10.0	620
W610 × 101	2,901	900	548	2.01	5.76	762
W410 × 132	2,868	889	552	2.68	9.20	541
W360 × 147	2,835	876	559	4.11	13.8	462
W530 × 109	2,819	874	537	1.95	5.85	666
W310 × 158	2,687	834	517	3.35	15.5	388
W460 × 113	2,671	828	519	2.81	8.26	554
W360 × 134	2,573	778	508	4.60	13.0	416
W310 × 143	2,409	747	466	3.32	14.2	347
W250 × 167	2,409	747	449	2.89	19.5	298
W460 × 106	2,393	743	451	1.83	5.97	487
W610 × 82	2,196	682	405	1.44	4.24	562
W530 × 74	1,803	560	336	1.40	4.15	410
W410 × 85	1,721	534	328	1.72	5.58	316
W460 × 74	1,655	514	316	1.78	5.15	333
W310 × 97	1,586	483	313	3.63	10.7	222
W530 × 66	1,563	485	290	1.36	3.96	351
W310 × 86	1,416	439	278	2.70	9.08	198
W310 × 79	1,277	396	251	2.67	8.60	177
W310 × 74	1,178	366	229	2.11	7.25	163
W200 × 100	1,149	357	216	2.28	14.5	113
W360 × 64	1,141	354	222	2.04	6.10	178
W460 × 52	1,090	338	205	1.31	3.75	212
W410 × 53	1,049	325	201	1.64	4.63	186

Table 4.1m *continued*

Shape	Z_x / 10^3 (mm^3)	ϕM_{px} (kN-m)	ϕM_{rx} (kN-m)	L_p (m)	L_r (m)	I_x / 10^6 (mm^4)
W250 × 73	990	308	194	2.73	9.63	113
W310 × 60	934	290	183	2.09	6.43	128
W310 × 52	839	260	163	1.66	5.06	119
W410 × 38.8	724	2251	137	1.21	3.41	125
W250 × 49.1	636	198	125	2.09	6.64	71.2
W310 × 38.7	610	190	119	1.62	4.54	84.9
W360 × 32.9	544	169	103	1.12	3.17	82.8
W250 × 32.7	426	132	83	1.43	4.21	49.1
W310 × 28.3	405	126	76	0.884	2.62	54.1
W200 × 35.9	379	117	74	1.73	5.76	34.4
W310 × 21	285	88.5	53	0.811	2.36	36.9
W200 × 26.6	279	86.5	54	1.32	4.11	25.8
W250 × 17.9	206	63.6	39	0.875	2.45	22.4
W200 × 15	145	44.6	28	0.957	2.60	12.8

Source: AISC Steel Construction Manual, 14th Edition

The bending coefficient is never less than 1.0. We can use this if we don't want to take the time to calculate it. Additionally, if a beam has several sections that are unbraced, then each section must be analyzed separately and the C_b value shall be calculated for each section.

Since the values in equation (4.8) are known for a given grade of steel, except C_b, we can plot strength as a function of unbraced length, like that in Figure 4.9. Such curves are available in the AISC *Steel Construction Manual*[4] for wide flange beams and channels. The tables assume C_b equals 1.0.

Regarding local buckling, we have only covered beams in the compact range. In the real world we seldom choose a non-compact or slender beam since it would not be very economical. Why use all that steel if you can't use it past the elastic range? In reality, virtually all of the rolled beam sections are compact, and local buckling is not a concern.

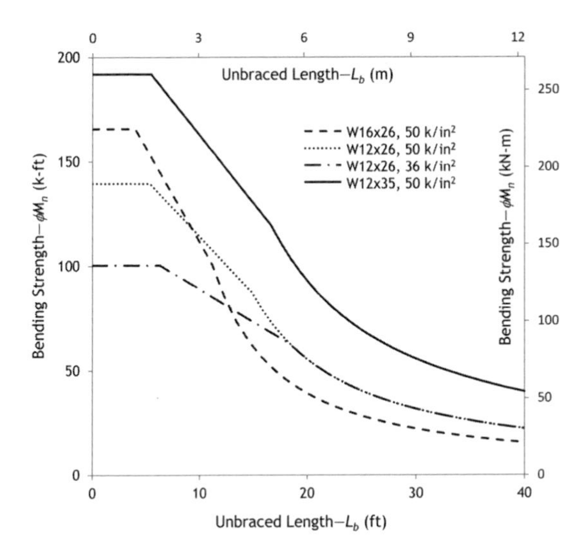

Figure 4.9 Bending strength as a function of unbraced length for two members and materials

4.2.3.1 Initial Beam Sizing

Using simple rules of thumb based on span to depth ratios, we can estimate beam depth for a given span. A quick way to do this is to make the depth in inches, equal to the span in feet divided by 2. For metric, multiply the span in meters by 40 to get the depth in millimeters. Multiply these by 1.5 for beams or **girders**. For example, a good starting point for a beam spanning 40 ft (12.2 m) is 20 in (500 mm).

Expanding on this, Table 4.2 provides initial member depths for a variety of members, based on span-depth ratios. To use the table, find your span along the top. Follow the column down to the row corresponding to your structural member type. Read off the depth in inches (millimeters).

If you know the bending moment on your beam, you can use Table 4.1 to select a bending member. For beams that are fully braced, use the ϕM_{px} column. This applies if your unbraced length

Table 4.2 Initial beam member sizing guide

SPAN DEPTH

Imperial Measures			*Span (ft)*											
		Span / Depth	*10*	*15*	*20*	*25*	*30*	*40*	*50*	*75*	*100*	*150*	*200*	*300*
System			*Depth (in)*											
Beam		**20**		9	12	15	18	24	30	45				
Composite Beam		**28**		8	9	11	13	17	21	32				
Crane Girder		**10**		18	24	30	36	48	60	90				
Floor Joist		**20**	6	9	12	15	18	24	30	45				
Roof Joist		**24**	6	8	10	13	15	20	25	38	50			
Plate Girder		**15**			16	20	24	32	40	60	80			
Truss		**12**			20	25	30	40	50	75	100	150	200	300
Space Frame		**16**					23	30	38	56	75	113	150	225

SPAN DEPTH

Metric Measures			*Span (m)*											
		Span / Depth	*3*	*4.5*	*6*	*7.5*	*9*	*12*	*15*	*23*	*30*	*45*	*60*	*90*
			Depth (mm)											
Beam		**20**		230	300	390	460	610	770	1,15	0			
Composite Beam		**28**		200	220	270	320	440	550	820				
Crane Girder		**10**		460	610	760	900	1,20	0	1,50	0			
Floor Joist		**20**	150	230	300	400	450	600	750	1,15	0			
Roof Joist		**24**	150	200	250	320	380	500	640	950	1,25			
Plate Girder		**15**			400	500	600	800	1,00	0	1,50	0	2,00	0
Truss		**12**			500	640	760	1,000	1,300	1,900	2,500	3,800	5,000	7,600
Space Frame		**16**					570	760	950	1,400	1,900	2,860	3,800	5,700

Notes

1) This table is for preliminary sizing only. Final section sizes must be calculated based on actual loading, length, and section size

2) Span ranges indicated are typical. Longer spans can be made with special consideration

L_b is less than the L_p column. If your brace spacing is more than this, but less than the L_r column, use ϕM_{rx} to estimate capacity. A much-expanded version of this table occurs in Part 3 of the *AISC Steel Manual.* Finally, we can use charts like Figure 4.9, again in Part 3, to determine bending capacity for any unbraced length.

Often in steel beam design, deflection controls our design. By rearranging the equations for deflection (found in Appendix 2), we can solve directly for the required moment of inertia I, then select the member from the section tables in Appendix 1. Table 4.3 provides these equations for common deflection criteria. (See section 4.4.1 for a discussion on deflection.)

4.3 DEMAND VERSUS CAPACITY

Once we have the nominal bending capacity M_n we multiply it by the strength reduction factor to get ϕM_n. We then compare it to the bending demand M_u. If $\phi M_n \geq M_u$, the member has adequate strength.

Previously, we looked at bending stress based on the magnitude of the moment in a given cross section (see section 4.2.1). Let's now look at the stresses in a simply supported and cantilever beam, and how they change along their length. For a simply supported beam, stress varies triangularly from compression at the top to tension at the bottom—zero at the middle—as illustrated in Figure 4.10a. A cantilever beam flips the stress direction upside down—tension at the top near the support, compression at the bottom—shown in Figure 4.10b.

We also know bending stress changes along the beam length, illustrated in Figure 4.11. In the single-span, simply supported beam case, the bending stress is zero at the ends, maximum and fully yielded at the middle, and somewhere between elastic and partially yielded in between. A cantilever is the opposite, with maximum stress at the supported end. A multi-span beam has positive bending stress at the mid-spans, but negative bending stress over the interior supports. Positive means tension bending stress at the bottom, negative indicates tension stress at the top.

4.4 SERVICEABILITY LIMIT STATES

Serviceability is based on day-to-day loading, deflection and vibration being the greatest concern. We don't want a floor to bounce like a

Table 4.3 Required moment of inertia solutions

Deflection Criteria	*Equation to Solve for Moment of Inertia I*	
	UNIFORM Load, Simple Support	
	Imperial (in)	*Metric (mm)*
1/600	$I_{req} = 1{,}125wL^3/E$	$I_{req} = 7.81 \times 10^9 wL^3/E$
1/480	$I_{req} = 900wL^3/E$	$I_{req} = 6.25 \times 10^9 wL^3/E$
1/360	$I_{req} = 675wL^3/E$	$I_{req} = 4.69 \times 10^9 wL^3/E$
1/240	$I_{req} = 450wL^3/E$	$I_{req} = 3.13 \times 10^9 wL^3/E$
1/180	$I_{req} = 338wL^3/E$	$I_{req} = 2.34 \times 10^9 wL^3/E$
1/120	$I_{req} = 225wL^3/E$	$I_{req} = 1.56 \times 10^9 wL^3/E$
	POINT Load, Simple Support	
	Imperial	*Metric*
1/600	$I_{req} = 1{,}800PL^2/E$	$I_{req} = 12.5 \times 10^9 PL^2/E$
1/480	$I_{req} = 1{,}440PL^2/E$	$I_{req} = 10.0 \times 10^9 PL^2/E$
1/360	$I_{req} = 1{,}080PL^2/E$	$I_{req} = 7.50 \times 10^9 PL^2/E$
1/240	$I_{req} = 720PL^2/E$	$I_{req} = 5.00 \times 10^9 PL^2/E$
1/180	$I_{req} = 540PL^2/E$	$I_{req} = 3.75 \times 10^9 PL^2/E$
1/120	$I_{req} = 360PL^2/E$	$I_{req} = 2.50 \times 10^9 PL^2/E$

Notes

1) Remember to use unfactored loads
2) Use the following units

Imperial	Metric
$w = k/ft$	$w = kN/m$
$P = k$	$P = kN$
$L = ft$	$L = m$
$E = k/in^2$	$E = MN/m^2$

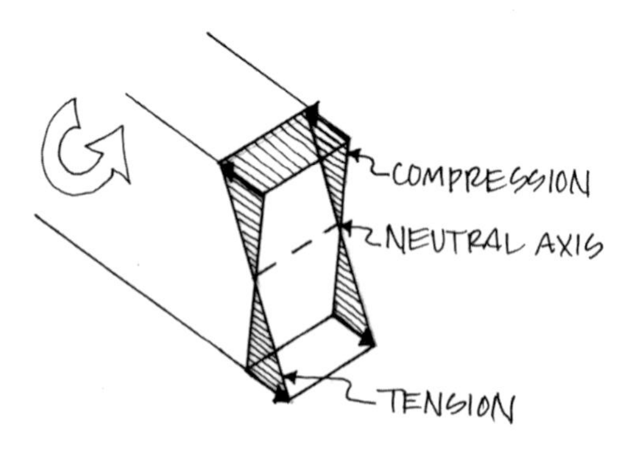

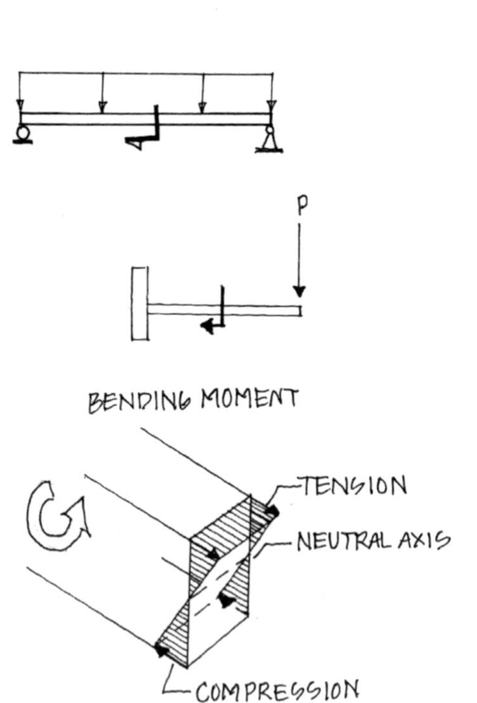

Figure 4.10 Bending stress distribution in (a) simple supported beam, (b) cantilever beam

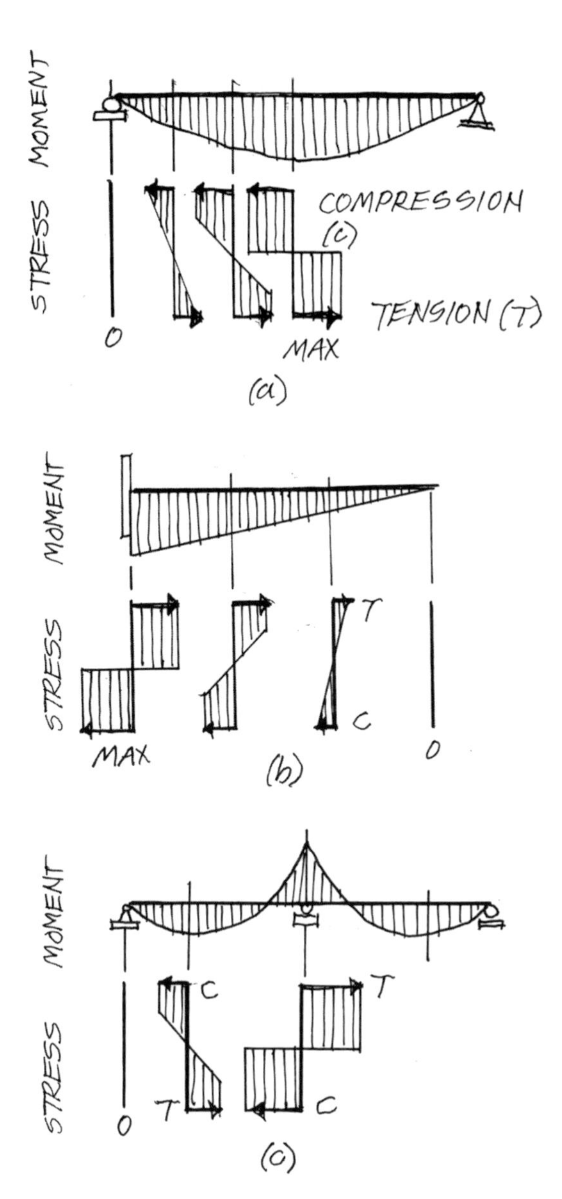

Figure 4.11 Bending stress variation along beam length for (a) single span, simple support, (b) cantilever, (c) multi-span

trampoline when we are the only one walking across it. But, if we were having a party and everyone was jumping with the beat, we would be willing to tolerate more movement.

We check serviceability limit states using ASD load combinations.

4.4.1 Deflection

We limit beam deflection according to the limits in Table 4.4, which are from *ASCE 7*.[1] These limits are given as a function of the span divided by a number. Table 4.5 applies them to various **span lengths**, providing allowable deflection. We can compare this directly to the deflection we calculate.

We commonly limit beam deflection to l/360 for **live load** and l/240 for total load. This means for a 15 ft (4.5 m) beam, the maximum deflection allowed when the live load is applied to the beam is

$$\delta_a = \frac{15\text{ ft}}{360}\frac{12\text{ in}}{1\text{ ft}} = 0.50\text{ in}$$

$$\delta_a = \frac{4{,}500\text{ mm}}{360} = 12.5\text{ mm}$$

Table 4.4 Deflection limits for beams

Member		*L or S*	
		or W	*D+L*
Roof Members	Plaster or Stucco Ceiling	L/360	L/240
	Non-Plaster Ceiling	L/240	L/180
	No Ceiling	L/180	L/120
Floor Members	Typical	L/360	L/240
	With Tile	L/480	L/360
	Supporting Masonry	L/600	L/480
Wall Members	Plaster or Stucco	L/360	–
	Other Brittle Finishes	L/240	–
	Flexible Finishes	L/120	–

Source: IBC 2012

l=Span, don't forget to convert to inches

Table 4.5 Calculated deflection limits for various spans

SPAN (L)

DEFLECTION

	Member Length in feet						
Limit	*15*	*20*	*25*	*30*	*35*	*40*	*50*
Criteria	*Allowable Deflection* δ_a *(in)*						
L/600	0.30	0.40	0.50	0.60	0.70	0.80	1.00
L/480	0.38	0.50	0.63	0.75	0.88	1.00	1.25
L/360	0.50	0.67	0.83	1.00	1.17	1.33	1.67
L/240	0.75	1.00	1.25	1.50	1.75	2.00	2.50
L/180	1.00	1.33	1.67	2.00	2.33	2.67	3.33
L/120	1.50	2.00	2.50	3.00	3.50	4.00	5.00

	Member Length in meters						
Limit	*4*	*6*	*8*	*9*	*10*	*12*	*15*
Criteria	*Allowable Deflection* δ_a *(mm)*						
L/600	6.7	10	13	15	17	20	25
L/480	8.3	13	17	19	21	25	31
L/360	11	17	22	25	28	33	42
L/240	17	25	33	38	42	50	63
L/180	22	33	44	50	56	67	83
L/120	33	50	67	75	83	100	125

while the maximum deflection allowed when the **dead** and live load are both applied to the beam is

$$\delta_a = \frac{15\text{ ft}}{240}\frac{12\text{ in}}{1\text{ ft}} = 0.75\text{ in}$$

$$\delta_a = \frac{4{,}500\text{ mm}}{240} = 18.8\text{ mm}$$

There are times we need a more stringent deflection criterion. If a beam supports masonry or tile veneer the total and live load deflections should be limited to l/600. If the beam supports a folding partition door, the deflections should be limited to the criteria obtained from the manufacturer.

Appendix 2 provides equations for calculating deflection under certain typical types of loading and support. These tables also contain equations for shears and moments. The equations give magnitudes but may require you to determine the direction and **sign convention** of the deflection or force. You can do this based on the applied load direction.

When sizing a member based on deflection, it is easier to take the deflection limit and work backwards to solve for the required moment of inertia. We have provided these equations in Table 4.3. We can then make our section from the tables in Appendix 1, or the *AISC Steel Construction Manual.* Section 3 of the *Manual* lists wide flange sections in order of decreasing moment of inertia, for easier beam selection.

4.4.2 Vibration

Vibration is another serviceability requirement that drives beam design. Design is comprised of dynamics and acceptance criteria based on how people perceive vibrations. We provide an introduction in *Special Structural Topics* in this series. If you want to go deeper, AISC has spent significant time to develop an industry standard for vibration design, known as *AISC Design Guide 11: Floor Vibrations due to Human Activity.*[2]

4.5 DESIGN STEPS

We are now ready to design beams, following these steps:

Step 1: Determine the structural layout
Step 2: Determine the loads
- a. Maximum service load
- b. Maximum factored load
- c. Determine ultimate moment diagram

Step 3: Determine material parameters
Step 4: Determine initial size
Step 5: Check all applicable strength limit states
- a. Yielding

b. Lateral torsional buckling
c. Local buckling

Step 6: Check member serviceability Limit States
a. Deflection
b. Vibration

Step 7: Summarize the final results

4.6 EXAMPLE: FLEXURAL MEMBER DESIGN

Step 1: Determine the structural layout

We will design the 30 ft (9 m) long steel beam shown in Figure 4.12. It has a 10 ft (3 m) **tributary width** and supports a floor with a dead and live **distributed** load and a concentrated point load at the center of the span, shown in Figure 4.13. Note the imperial inputs don't directly convert to the metric values used.

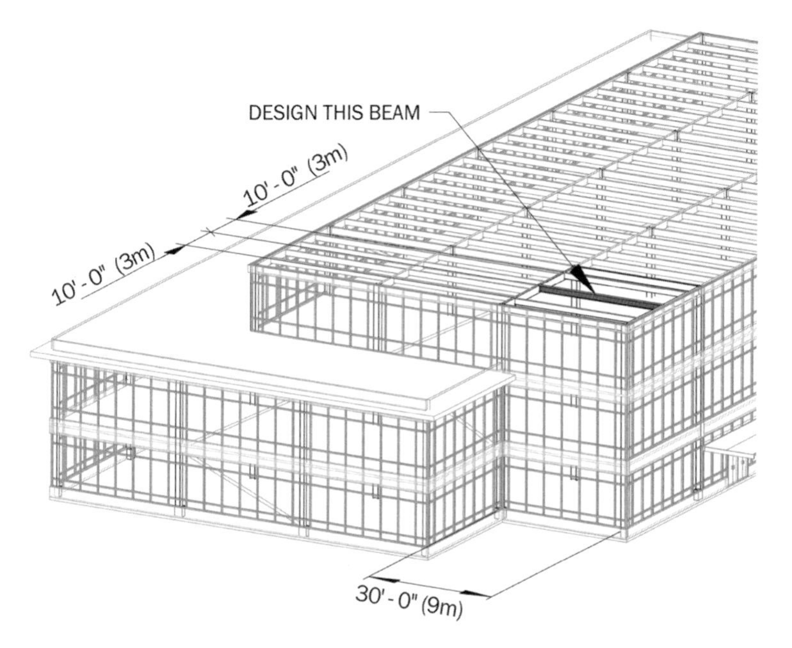

Figure 4.12 Design example layout

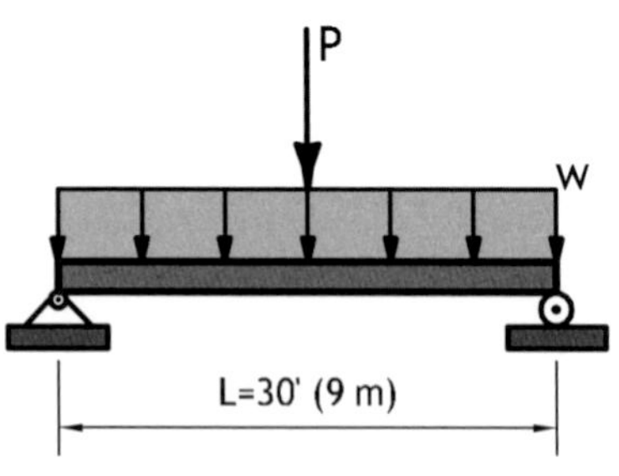

Figure 4.13 Beam example free body diagram

We will assume the floor deck is not welded down to the beam and thus does not laterally brace it. Instead, beams at the third points, shown in Figure 4.14, provide lateral stability bracing.

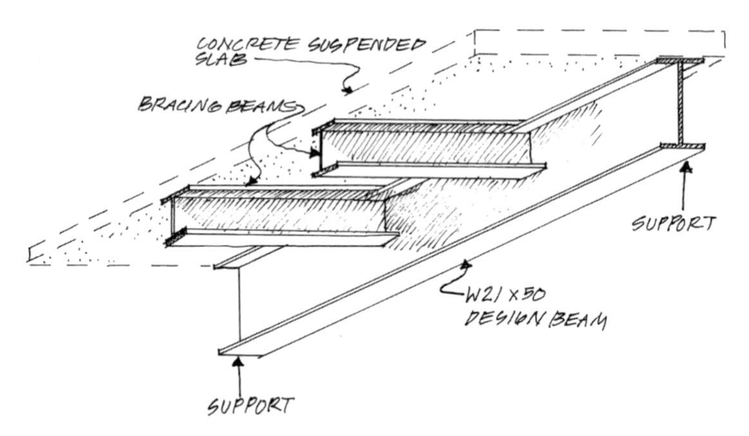

Figure 4.14 View of bracing for example

Step 2: Determine the loads

The unit loads at the floor are as follows:

$q_D = 65\ \text{lb/ft}^2$	$q_D = 3.1\ \text{kN/m}^2$
$q_L = 100\ \text{lb/ft}^2$	$q_L = 4.8\ \text{kN/m}^2$

The point loads at the floor are:

$P_D = 2.0\ \text{k}$	$P_D = 10.0\ \text{kN}$
$P_L = 4.0\ \text{k}$	$P_L = 20.0\ \text{kN}$

Looking at the **free body** diagram of the beam in Figure 4.13, we see the loads we need to calculate and the support conditions. We begin by finding the service and factored load combinations, which we will use for deflection and strength calculations, respectively.

2a Maximum Service Load

The load combination D + L will control, since this is a gravity floor beam.

$$w = [q_D + q_L]L_t$$

$$w = [65 \text{ lb/ft}^2 + 100 \text{ lb/ft}^2]10 \text{ ft}\left(\frac{1 \text{ k}}{1{,}000 \text{ lb}}\right)$$
$$= 1.65 \text{ k/ft}$$

$$w = [3.1 \text{ kN/m}^2 + 4.8 \text{ kN/m}^2]3 \text{ m}$$
$$= 23.7 \text{ kN/m}$$

$$P = P_D + P_L$$

$$P = 2.0 \text{ k} + 4.0 \text{ k}$$
$$= 6.0 \text{ k}$$

$$P = 10 \text{ kN} + 20 \text{ kN}$$
$$= 30.0 \text{ kN}$$

2b Maximum Factored Load

The factored load combination is 1.2D + 1.6L

$$w_u = [1.2q_D + 1.6q_L]L_t$$

$$w_u = [1.2(65 \text{ lb/ft}^2) + 1.6(100 \text{ lb/ft}^2)]10.0 \text{ ft}\left(\frac{1 \text{ k}}{1{,}000 \text{ lb}}\right)$$
$$= 2.38 \text{ k/ft}$$

$$w_u = [1.2(3.1 \text{ kN/m}^2) + 1.6(4.8 \text{ kN/m}^2)]3 \text{ m}$$
$$= 34.2 \text{ kN/m}$$

$$P_u = 1.2P_D + 1.6P_L$$

$$P_u = 1.2(2.0 \text{ k}) + 1.6(4.0 \text{ k})$$
$$= 8.8 \text{ k}$$

$$P_u = 1.2(10.0 \text{ kN}) + 1.6(20.0 \text{ kN})$$
$$= 44.0 \text{ kN}$$

2c Determine Factored Moment Diagram

Based on equations in Appendix 2

$$(M_u)_x = \frac{wx(L-x)}{2} + \frac{Px}{2} \quad \text{for } x \le 1/2$$

$$(M_u)_x = \frac{wx(L-x)}{2} + \frac{P(L-x)}{2} \quad \text{for } x > 1/2$$

Table 4.6 provides the moment at key points from the left support. This will give us our maximum demand, and values needed to calculate C_b.

Step 3: Determine Material Parameters

We will use a W section for the beam, made of A992 steel. The yield strength, and elastic modulus are

F_y = 50 k/in^2 (345 N/mm^2)
E = 29,000 k/in^2 (200,000 N/mm^2).

Step 4: Determine Initial Size

Taking half the span in inches, times 1.5 because of the heavy load, we get a 22.5 in (570 mm). Let's try a W21 × 50 (W530 × 74) and see where that gets us. Section properties that we will need are (from Table A1.1)

b_f = 6.53 in
t_f = 0.535 in
Z_x = 110 in^3
S_x = 94.5 in^3
I_x = 984 in^4

b_f = 166 mm
t_f = 13.6 mm
Z_x = 1,800 × 10^3 mm^3
S_x = 1,550 × 10^3 mm^3
I_x = 410 × 10^6 mm^4

Let's now check local buckling and make sure the flange is compact.

The width to thickness ratio is

$$\frac{b}{t} = \frac{b_f}{2t_f}$$

$$= \frac{6.53 \text{ in}}{2(0.535 \text{ in})} = 6.10$$

$$= \frac{166 \text{ mm}}{2(13.6 \text{ mm})} = 6.10$$

We compare this to the limiting width to thickness ratio λ_p

Table 4.6 Moments at various points along the beam length

Distance from Support (x), ft	0.0	2.5	5.0	7.5	10.0	12.5	15.0	17.5	20.0	22.5	25.0	27.5	30.0
Moment (M_u), k-ft	0.0	92.8	170.8	233.8	282.0	315.3	333.8	315.3	282.0	233.8	170.8	92.8	0.0
	↑ Support		Unbraced Section 1		⫳ Brace Point		Unbraced Section 2		⫳ Brace Point		Unbraced Section 3		↑ Support
Distance from Support (x), m	0.0	0.8	1.5	2.3	3.0	3.8	4.5	5.3	6.0	6.8	7.5	8.3	9.0
Moment (M_u), kN-m	0.0	122.3	225.4	309.2	373.8	419.2	445.3	419.2	373.8	309.2	225.4	122.3	0.0
	↑ Support		Unbraced Section 1		⫳ Brace Point		Unbraced Section 2		⫳ Brace Point		Unbraced Section 3		↑ Support

$$\lambda_p = 0.38\sqrt{\frac{E}{F_y}}$$

$$= 0.38\sqrt{\frac{29{,}000\ \text{k/in}^2}{50\ \text{k/in}^2}} = 9.15$$

$$= 0.38\sqrt{\frac{200{,}000\ \text{N/mm}^2}{345\ \text{N/mm}^2}} = 9.15$$

Because *b/t* is less than this value, we know our flange is compact and can take substantial yielding before locally buckling, and that we don't need to use different equations below.

Step 5: Check all Applicable Strength Limit States

We begin by checking yielding

$$M_P = F_y Z_x = 50\ \text{k/in}^2\left(110\ \text{in}^3\right)\frac{1\ \text{ft}}{12\ \text{in}}$$
$$= 458\ \text{k} - \text{ft}$$

$$= 345\ \text{N/mm}^2\left(1{,}800\times 10^3\ \text{mm}^3\right)\left(\frac{1\ \text{m}}{1000\ \text{mm}}\right)\left(\frac{1\ \text{kN}}{1000\ \text{N}}\right)$$
$$= 621\ \text{kN} - \text{m}$$

This is larger than the demand, so we know we are heading in the right direction.

We next check lateral torsional buckling. The unbraced beam length is one third of the span, shown in Figure 4.14, as follows

$L_b = 10.0$ ft $\qquad L_b = 3.0$ m

We will need L_r and L_p, using equations 4.4 and 4.5 and additional section properties from the *Steel Manual*. Alternatively, we can pull them from Table 4.1.

$L_p = 4.59$ ft $\qquad L_p = 1.40$ m

$L_r = 13.6$ ft $\qquad L_r = 4.15$ m

Because L_b is between these, we know the beam will be in the inelastic lateral torsional buckling range. Pretty exciting!

Next, we determine the lateral torsional buckling factor C_b for each section. While we can conservatively take C_b=1.0, it may under predict the moment capacity. Let's see how much of a difference it makes. For Spans 1 and 3, and using the moments from step 2 (with units of k-ft and kN-m), we have:

$$C_b = \frac{12.5M_{max}}{2.5M_{max} + 3M_A + 4M_B + 3M_C}$$
$$= \frac{12.5(282)}{2.5(282) + 3(92.8) + 4(171) + 3(234)}$$
$$= 1.49$$

$$= \frac{12.5(374)}{2.5(374) + 3(122) + 4(225) + 3(309)}$$
$$= 1.49$$

And now for the middle span,

$$C_b = \frac{12.5M_{max}}{2.5M_{max} + 3M_A + 4M_B + 3M_C}$$
$$= \frac{12.5(334)}{2.5(334) + 3(315) + 4(334) + 3(315)}$$
$$= 1.03$$

$$= \frac{12.5(445)}{2.5(445) + 3(419) + 4(445) + 3(419)}$$
$$= 1.03$$

We see for the middle span, C_b is almost 1, so we don't get much of an advantage. Now we know.

Now to find the nominal bending capacity, considering LTB, we check section 2, where the demand is greatest and C_b is lowest.

$$M_n = C_b\left[M_p - (M_p - 0.7F_yS_x)\left(\frac{L_b - L_p}{L_r - L_p}\right)\right] \le M_p$$

$$M_n = 1.03\left[458 \text{ k ft} - \left(458 \text{ k ft} - 0.7\left(\frac{50 \text{ k}}{\text{in}^2}\right)(94.5 \text{ in}^3)\left(\frac{1 \text{ ft}}{12 \text{ in}}\right)\right)\left(\frac{10.0 \text{ ft} - 4.59 \text{ ft}}{13.56 \text{ ft} - 4.59 \text{ ft}}\right)\right]$$
$$= 359 \text{ k ft}$$

$$M_n = 1.03\left[621 \text{ N m} - \left(621 \text{ kN m} - 0.7\left(\frac{345 \text{ N}}{\text{mm}^2}\right)(1.55\times10^6 \text{ mm}^3)\left(\frac{1 \text{ kN m}}{1\times10^6 \text{ N mm}}\right)\right)\left(\frac{3.0 \text{ m} - 1.4 \text{ m}}{4.15 \text{ m} - 1.4 \text{ m}}\right)\right]$$
$$= 492 \text{ N m}$$

This is below the plastic moment capacity M_p. Multiplying by ϕ we get the capacity

$$\phi M_n = 0.9(359 \text{ k ft})$$
$$= 323 \text{ k ft}$$

$$\phi M_n = 0.9(492 \text{ kN m})$$
$$= 443 \text{ kN m}$$

Since the capacity is slightly less than the demand, our beam does not work. However, since they are so close, we can bump up a size in Table A1.1 to a W21 × 73 (W530 × 109), and know we are OK.

Step 6: Check Member Serviceability Limit States

Next, let's check the deflection of the W21 × 50 (W530 × 74), to see if it works for deflection. For live loads,

$$w_L = q_L l_t$$
$$= 0.10\ \text{k/ft}^2(10\ \text{ft})$$
$$= 1.0\ \text{k/ft}$$

$$= 4.8\ \text{kN/m}^2(3\ \text{m})$$
$$= 14.4\ \text{kN/m}$$

$$\delta_L = \frac{5w_L L^4}{384EI} + \frac{PL^3}{48EI}$$

$$\delta_L = \frac{5(1.00\ \text{k/ft})(30\ \text{ft})^4}{384(29\times10^3\ \text{k/in}^2)(984\ \text{in}^4)}\left(\frac{12\ \text{in}}{1\ \text{ft}}\right)^3 + \frac{4.0\ \text{k}(30\ \text{ft})^3}{48(29\times10^3\ \text{k/in}^2)(984\ \text{in}^4)}\left(\frac{12\ \text{in}}{1\ \text{ft}}\right)^3$$
$$= 0.775\ \text{in}$$

$$\delta_L = \frac{5(14.4\ \text{kN/m})(9\ \text{m})^4}{384(200\ \text{kN/mm}^2)(410\times10^6\ \text{mm}^4)}\left(\frac{1{,}000\ \text{mm}}{1\ \text{m}}\right)^3 + \frac{20\ \text{kN}(90\ \text{m})^3}{48(200\ \text{kN/mm}^2)(410\times10^6\ \text{mm}^4)}\left(\frac{1{,}000\ \text{mm}}{1\ \text{m}}\right)^3$$
$$= 18.7\ \text{mm}$$

$$\delta_{a,L} = \frac{L}{360}$$

$$= \frac{30\ \text{ft}}{360}\left(\frac{12\ \text{in}}{1\ \text{ft}}\right)^3 = 1.00\ \text{in}$$

$$= \frac{9\ \text{m}}{360}\left(\frac{1{,}000\ \text{mm}}{1\ \text{m}}\right)^3 = 25\ \text{mm}$$

Because the actual deflection is less than the allowable, it appears our original beam works for live load deflection. On your own, check the total load for a deflection criterion of L/240 for our larger beam.

Step 7: Summarize the Final Results

The final beam design is a W21 × 73 (W530 × 109).

4.7 WHERE WE GO FROM HERE

There are several areas of design that are not covered in depth in this chapter. First are the vibration requirements for beams found in *AISC Design Guide 11*[2]*: Vibrations of Steel-Framed Structural Systems Due to Human Activity*. Vibrations can be an important design requirement for everyday use, and critical for sensitive equipment. Chapter 1 of *Special Structural Topics*[3] provides a sound introduction to vibration.

Another area of further study is the local buckling limiting slenderness parameters, λ_p and λ_r, for different sections such as built up sections, round hollow structural sections, or angles. These are described in section B of the AISC 360 specification.

While design of elements which fall in the slenderness range for local buckling is not encouraged, there are situations where the capacity of slender elements may be required. AISC 360 section F provides design equations for determining the local buckling capacities for beams which fall within the slenderness range.

NOTES

1. *Design Guide 11: Floor Vibrations Due To Human Activity*, AISC/CISC Steel Design Guide Series 11, American Institute of Steel Construction, 1997.
2. Esra Hasanbas Persellin. "Vibration." In *Special Structural Topics*, edited by Paul W. McMullin and Jonathan S. Price. (New York: Routledge, 2017).
3. Seaburg, P.A., Carter, C.J., *Torsional Analysis of Structural Steel Members*, Steel Design Guide 9 (Chicago: American Institute of Steel Construction, 1997).

Steel Shear and Torsion

Chapter 5

Paul W. McMullin

Shear is fundamental to how bending members resist load. It holds the layers of a member together, causing them to act as one. This chapter will focus on how we size beams for shear force. We will look at the direct shear method and introduce **tension field** action.

Let's take a moment and conceptually understand the fundamentals of shear behavior. You are likely familiar with the action scissors make when cutting paper or fabric. The blades are perpendicular to the material, going in opposite directions. This creates a tearing of the material like that shown in Figure 5.1. In beam shear, the action is similar, but the movement of material is parallel to the length of the beam. The top portion moves relative to the bottom, illustrated in Figure 5.2. Shear stress is zero at the top and bottom edge, and a maximum at the middle, shown in Figure 5.3.

Shear strength is fundamentally tied to bending strength and stiffness. If we take a stack of paper and lay it across two supports, it sags (Figure 5.4a), unable to carry even its own load. If we glue each strip of paper together, we get a beam with enough strength and stiffness to carry a reasonable load, as shown in Figure 5.4b.

5.1 STABILITY

Web buckling can occur when it becomes too slender, creating a reduction in strength. This is handled in two ways, direct method and tension field action. For direct shear, we multiply the shear strength by the web shear coefficient C_v, discussed in the next section. This keeps the stresses low enough to avoid buckling the web.

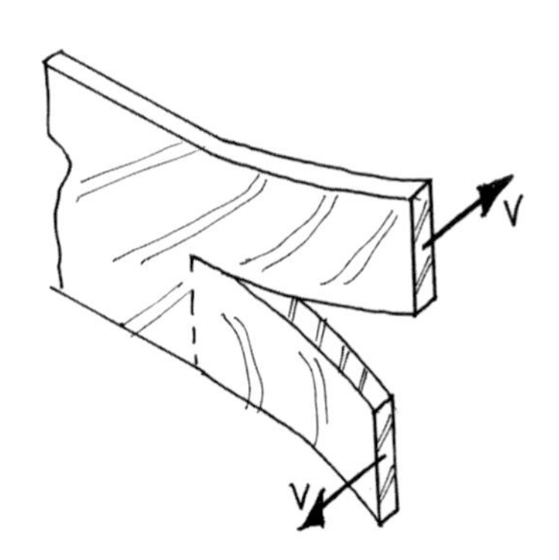

Figure 5.1 Shearing action similar to scissors

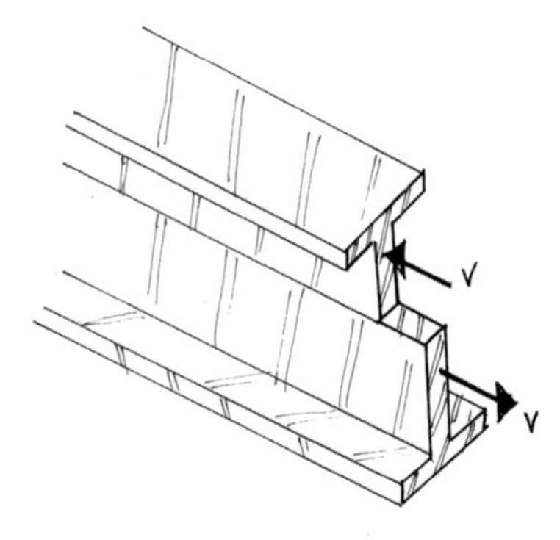

Figure 5.2 Shearing action from bending

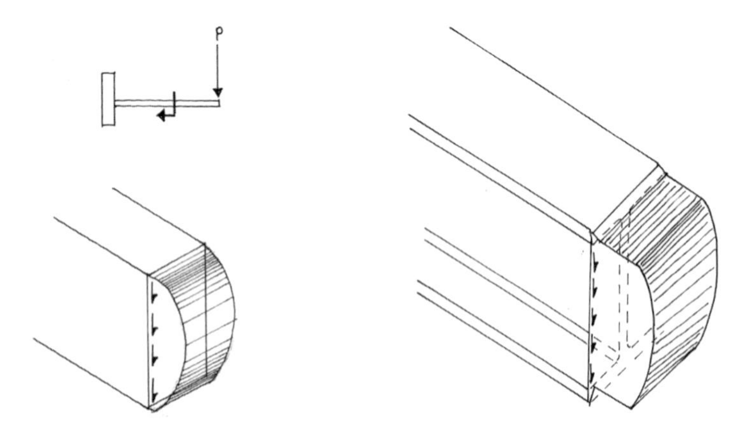

Figure 5.3 Shear stress distribution in (a) rectangular cross-section and (b) wide flange

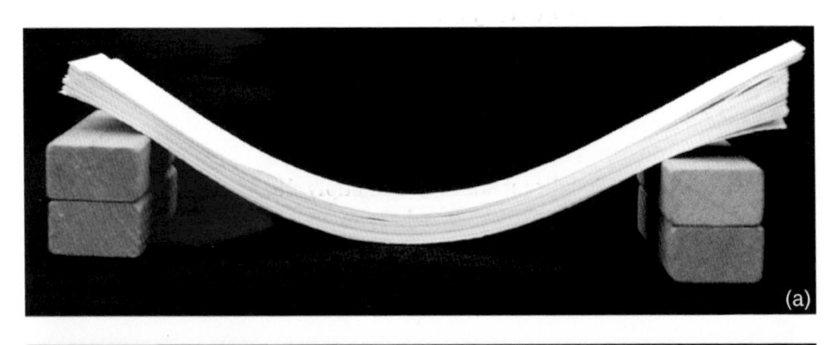

Figure 5.4 Paper beam with layers (a) unglued and (b) glued Lego™ figures courtesy Peter McMullin

In tension field action, the web buckles, but we add web **stiffeners** to prevent collapse. To understand this, we look at the shear force on a small portion of the web, illustrated in Figure 5.5a. The shear forces are parallel to the applied load, and balance each other. Turning the force block 45 degrees, the forces resolve into compression, causing the web to buckle, and tension, providing strength—opposite of what we want in concrete. Figure 5.6 shows this phenomenon in a test specimen.

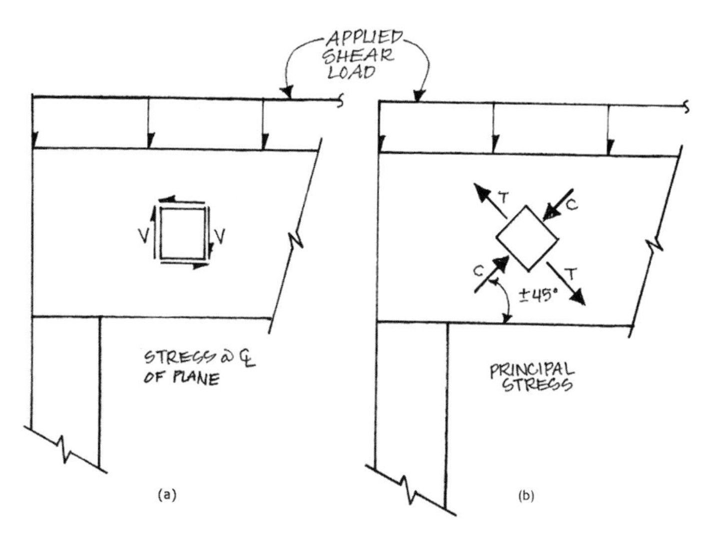

Figure 5.5 (a) Shear force in a beam web and (b) equivalent tension and compression forces

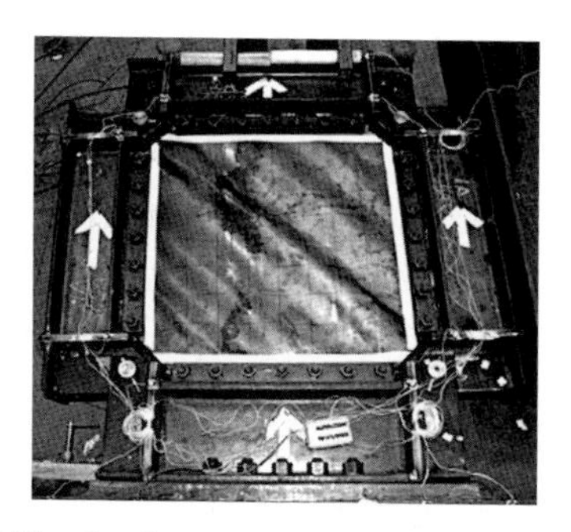

Figure 5.6 Tension field action in test specimen

Reprinted with permission, *Structure* magazine, September 2014

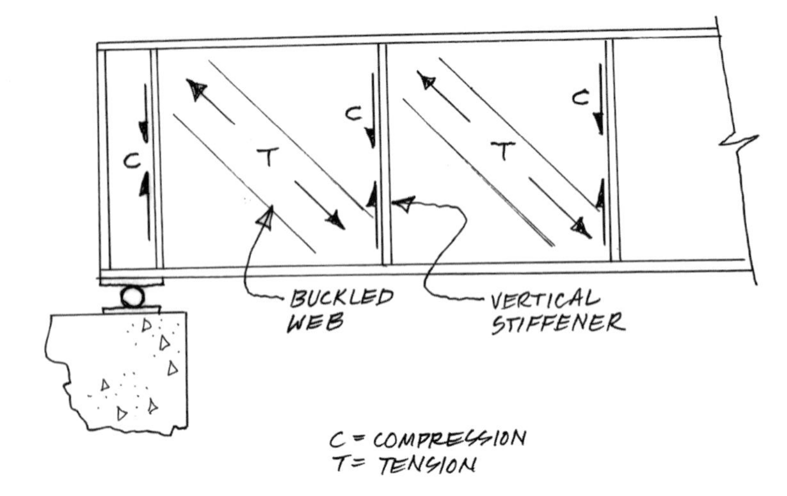

Figure 5.7 Tension field pseudo truss

Tension field action creates a pseudo truss. The buckled web creates diagonal tension members, while the stiffeners provide vertical compression elements, illustrated in Figure 5.7. This approach is highly efficient and often used in deep highway girders, where the webs are slender. Sometimes the sun hits these girders in a way that you can see a slight ripple in the web.

5.2 CAPACITY

5.2.1 Direct Shear

In the direct shear method, we proportion the web to carry the forces without buckling. Nominal shear strength V_n is given by

$$V_n = 0.6F_y A_w C_v \tag{5.1}$$

where

F_y = tension yield strength, lb/in² (MN/m²)
A_w = web area = dt_w
C_v = web shear coefficient.

The 0.6 accounts for the fact that shear yield occurs at approximately 60% of the tension yield strength. The web area is taken as the thickness of the web times the full member depth. The web shear coefficient depends on the slenderness, as follows:

For rolled I-shaped members:

$$\frac{h}{t_w} \le 2.24\sqrt{\frac{E}{F_y}}$$
$$C_v = 1.0 \tag{5.2}$$

For all other shapes, except round HSS:

When $\dfrac{h}{t_w} \le 1.10\sqrt{\dfrac{k_v E}{F_y}}$

$$C_v = 1.0 \tag{5.3}$$

$$\frac{h}{t_w} \ge 1.10\sqrt{k_v E/F_y}$$
$$C_v = \frac{1.10\sqrt{k_v E/F_y}}{h/t_w} \tag{5.4}$$

where

h = clear distance between flanges, in (mm), shown in Figure 5.8.
t_w = web thickness, in (mm)
E = modulus of elasticity, 29,000 k/in^2 (200 GN/m^2)
F_y = tension yield stress, k/in^2 (MN/m^2)
k_v = 5 for h/t_w<260
= 10 for web stiffeners spaced at h
= 25 for web stiffeners spaced at $h/2$.

The strength reduction factor for direct shear ϕ is 0.9, except for rolled I-shapes, when

$\dfrac{h}{t_w} \le 2.24\sqrt{\dfrac{E}{F_y}}$, it is 1.0.

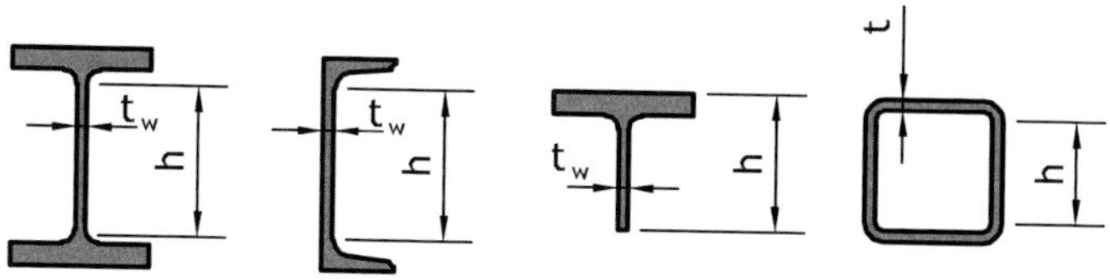

Figure 5.8 Web height and thickness definitions

Interestingly, most rolled shapes in the *Steel Manual* have compact webs with $C_v = 1.0$.

5.2.2 Torsion

Torsional moments occur when loads are eccentric to a members shear center. For wide flange and hollow structural shapes, this is the centerline of the member, for channels it is away from the web, and for angles at the centerline of the leg parallel to the force, illustrated in Figure 5.9.

When the load is not applied to the shear center, the member will twist, like the channel show in Figure 5.10.

For open shapes like wide flanges and channels, it is best to avoid torsion, as they have low torsion strength. Closed shapes, like HSS, carry torsion efficiently. We will look at the strength equations for rectangular HSS. For open shapes, see *AISC Design Guide 9.* [1] The nominal torsion capacity T_n for HSS is given by Equation 5.5.

$$T_n = F_{cr} C \tag{5.5}$$

where

F_{cr} = critical torsion stress, lb/in² (MN/m²)
C = torsion parameter, in³ (mm³), see Table A1.4

The critical torsion stress is a function of wall slenderness (Equation 5.6).

$$\text{When } \frac{h}{t} \le 2.45\sqrt{\frac{E}{F_y}}$$

$$F_{cr} = 0.6F_y \tag{5.6}$$

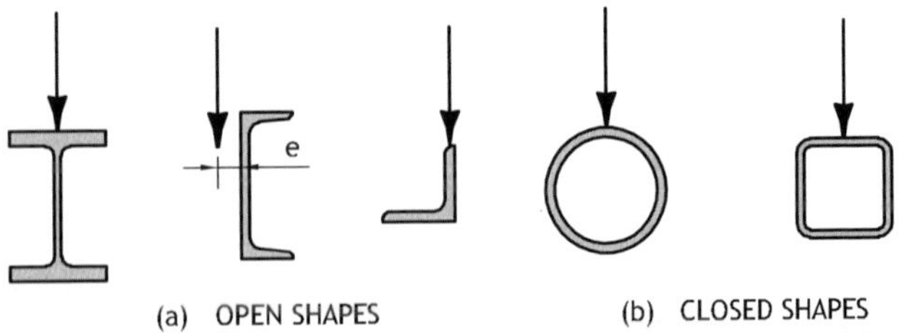

Figure 5.9 Shear center for various shapes, indicated by arrow

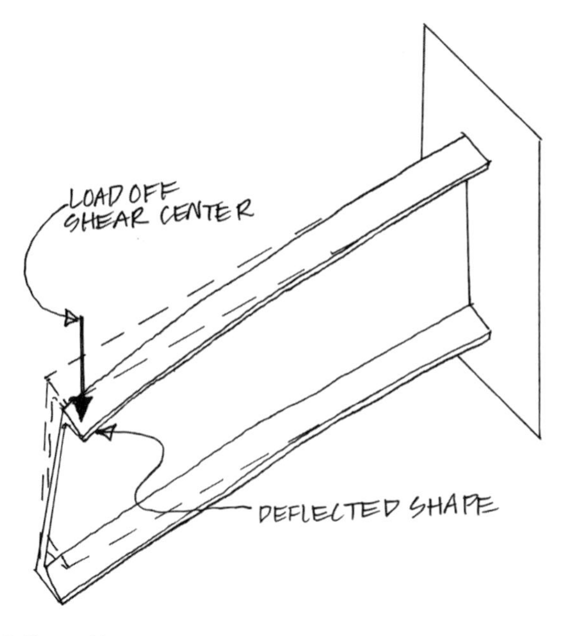

Figure 5.10 Channel loaded off shear center, showing deflection

When $2.45\sqrt{E/F_y} < h/t \le 3.07\sqrt{E/F_y}$

$$F_{cr} = \frac{0.06F_y(2.45)\sqrt{E/F_y}}{h/t}$$

When $3.07\sqrt{E/F_y} < h/t < 260$

$$F_{cr} = \frac{0.458\pi^2 E}{(h/t)^2}$$

The strength reduction factor for torsion ϕ is 0.9.

Initial Sizing

To help you select a beam large enough to carry a given shear load V_u, Table 5.1 provides the shear strength ϕV_n for various beam sizes and web areas.

Table 5.1 Shear strength for various beam sizes and web areas

Imperial			*Metric*		
Shape	A_w *(in²)*	ϕV_{nx} *(k)*	*Shape*	A_w *(mm²)*	ϕV_{nx} *(kN)*
W36 × 652	81.0	2,430	W920 × 970	52,237	10,809
W40 × 593	77.0	2,310	W1000 × 883	49,658	10,275
W14 × 730	68.8	2,060	W360 × 1086	44,366	9,163
W27 × 539	64.0	1,920	W690 × 802	41,306	8,541
W40 × 392	59.1	1,770	W1000 × 584	38,111	7,873
W33 × 387	45.4	1,360	W840 × 576	29,26	6,050
W44 × 335	45.3	1,360	W1100 × 499	29,239	6,050
W30 × 391	45.2	1,350	W760 × 582	29,130	6,005
W24 × 370	42.6	1,280	W610 × 551	27,458	5,694
W18 × 311	33.9	1,020	W460 × 464	21,868	4,537
W36 × 210	30.5	914	W920 × 313	19,652	4,066
W12 × 336	29.9	897	W310 × 500	19,293	3,990
W33 × 201	24.1	723	W840 × 299	15,545	3,216
W40 × 149	24.1	650	W1000 × 222	15,526	2,891
W30 × 211	23.9	718	W760 × 314	15,450	3,194
W27 × 217	23.6	707	W690 × 323	15,208	3,145
W36 × 135	21.4	577	W920 × 201	13,781	2,567
W21 × 201	20.9	628	W530 × 300	13,503	2,793
W12 × 230	19.5	584	W310 × 342	12,567	2,598
W14 × 257	19.4	581	W360 × 382	12,485	2,584
W33 × 118	18.1	489	W840 × 176	11,674	2,175
W24 × 162	17.6	529	W610 × 241	11,371	2,353
W30 × 116	17.0	509	W760 × 173	10,968	2,264
W18 × 158	16.0	479	W460 × 235	10,295	2,131
W30 × 90	13.9	374	W760 × 134	8,945	1,664

Table 5.1 *continued*

Imperial			*Metric*		
Shape	A_w *(in²)*	ϕV_{nx} *(k)*	*Shape*	A_w *(mm²)*	ϕV_{nx} *(kN)*
W24 × 103	13.5	404.0	W610 × 153	8,694	1,797
W21 × 122	13.0	391.0	W530 × 182	8,400	1,739
W27 × 84	12.3	368.0	W690 × 125	7,924	1,637
W12 × 152	11.9	358.0	W310 × 226	7,690	1,59
W16 × 100	9.95	298.0	W410 × 149	6,416	1,326
W24 × 55	9.32	252.0	W610 × 82	6,014	1,121
W18 × 71	9.16	275.0	W460 × 106	5,908	1,223
W10 × 112	8.61	258.0	W250 × 167	5,553	1,148
W21 × 57	8.55	256.0	W530 × 85	5,513	1,139
W21 × 44	7.25	217.0	W530 × 66	4,674	965
W16 × 57	7.05	212.0	W410 × 85	4,550	943
W12 × 96	6.99	210.0	W310 × 143	4,506	934
W18 × 50	6.39	192.0	W460 × 74	4,123	854
W14 × 90	6.16	185.0	W360 × 134	3,974	823
W18 × 35	5.31	159.0	W460 × 52	3,426	707
W8 × 67	5.13	154.0	W200 × 100	3,310	685
W12 × 65	4.72	142.0	W310 × 97	3,045	632
W16 × 36	4.69	141.0	W410 × 53	3,026	627
W12 × 50	4.51	135.0	W310 × 74	2,912	601
W12 × 58	4.39	132.0	W310 × 86	2,834	587
W14 × 43	4.18	125.0	W360 × 64	2,696	556
W12 × 53	4.17	125.0	W310 × 79	2,693	556
W16 × 26	3.93	106.0	W410 × 38.8	2,532	472
W12 × 35	3.75	113.0	W310 × 52	2,419	503
W12 × 40	3.51	105.0	W310 × 60	2,265	467

Table 5.1 *continued*

Imperial			*Metric*		
Shape	A_w *(in²)*	ϕV_{nx} *(k)*	*Shape*	A_w *(mm²)*	ϕV_{nx} *(kN)*
W10 × 49	3.40	102	W250 × 73	2,194	454
W14 × 22	3.15	94.5	W360 × 32.9	2,033	420
W12 × 19	2.87	86.0	W310 × 28.3	1,850	383
W10 × 33	2.82	84.7	W250 × 49.1	1,820	377
W12 × 26	2.81	84.2	W310 × 38.7	1,810	375
W10 × 22	2.45	73.4	W250 × 32.7	1,579	326
W12 × 14	2.38	64.3	W310 × 21	1,535	286
W8 × 24	1.94	58.3	W200 × 35.9	1,253	259
W10 × 12	1.88	56.3	W250 × 17.9	1,210	250
W8 × 18	1.87	56.2	W200 × 26.6	1,208	250
W8 × 10	1.34	40.2	W200 × 15	865	179

Source: AISC Steel Construction Manual, 14th Edition

5.3 DEMAND VS CAPACITY

5.3.1 Direct Shear

Once we have the shear capacity ϕV_n we compare it to the shear demand V_u. If $\phi V_n \geq V_u$, the member has adequate shear capacity.

The shear force distribution in a beam depends on support conditions and loading. In a single-span, simply supported beam, with a uniform distributed load, the shear stress is zero at the middle and maximum at the ends. A cantilever is the opposite, with maximum force at the supported end. A multi-span beam has maximum shear at the supports. These are illustrated in Figure 5.11.

5.3.2 Torsion

Similar to direct shear, once we calculate the torsion capacity $\phi_T T_n$, we compare it to the shear demand T_u. If $\phi_T T_n \geq T_u$, the member has adequate torsion capacity.

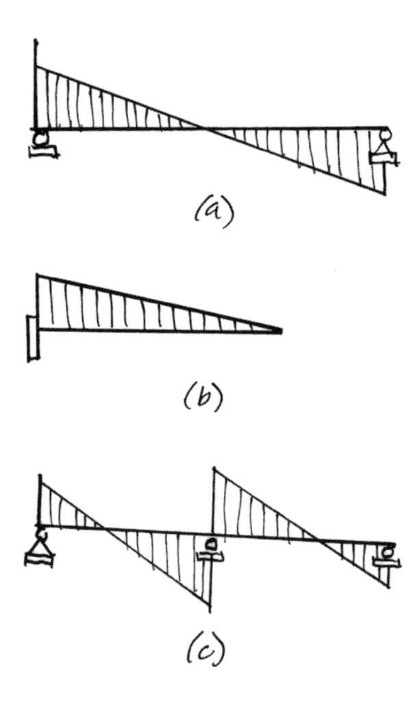

Figure 5.11 Shear force variation at points along (a) simply supported, (b) cantilever, and (c) multi-span beams

5.4 DEFLECTION

Shear action contributes little to bending deflection—usually only 3–5%— and can be ignored. Refer to Chapter 4 for additional discussion on deflections calculations and acceptance criteria.

Torsion deflection, on the other hand, can have a substantial effect on deflections, particularly when other members cantilever from torsion members.

5.5 DETAILING

Detailing considerations for shear are similar to those found in Chapter 8 for connections. Keep in mind shear force is a maximum near supports and at the mid-height of the cross-section. Any field modifications for piping or conduit in these areas will reduce shear strength and should be carefully evaluated by a professional engineer.

5.6 DESIGN STEPS

Step 1: Draw the structural layout, include span dimensions and tributary width

Step 2: Determine Loads
- Unit Loads
- Load Combinations yielding a line load
- Member maximum shear demand

Step 3: Identify Material Parameters

Step 4: Estimate initial size or use size from bending and deflection check

Step 5: Calculate capacity and compare to demand

Step 6: Summarize the results

5.7 DESIGN EXAMPLE

5.7.1 Direct Shear Example

Building on the beam example in Chapter 4, we will now check the shear capacity of the wide flange beam where the span has been doubled, due to a column interference, as shown in Figure 5.12. This gives us a beam span L of 60 ft (18.3 m).

See the example in Chapter 4 for steps 1 to 2b. In this case, we don't have the point loads on our beam.

Step 2c: Determine Member Shear Demand

We are only concerned with the maximum shear, which occurs at the ends. Using the formulas in Appendix 2, and $w_u = 2.38$ k/ft (34.2 kN/m), we see half the total beam load is supported at each end.

$$V_u = w_u \frac{L}{2}$$

$$= 2.38 \text{ k/ft}\left(\frac{60 \text{ ft}}{2}\right)$$

$$= 71.4 \text{ k}$$

$$= 34.2 \text{ kN/m}\left(\frac{18.3 \text{ m}}{2}\right)$$

$$= 313 \text{ kN}$$

Step 3: Material Parameters

Because we are a using a wide flange beam, our yield stress is

$$F_y = 50 \text{ k/in}^2$$

$$F_y = 345 \text{ MN/m}^2$$

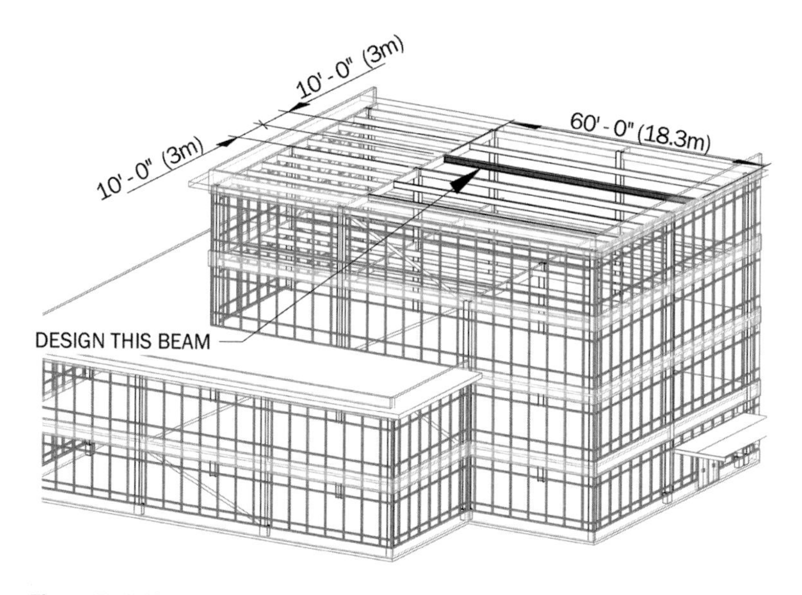

Figure 5.12 Shear example layout

Step 4: Initial Size

Choosing a W30 × 116 (W760 × 173), which is reasonable for bending, we know

$$d = 30 \text{ in} \qquad d = 762 \text{ mm}$$
$$t_w = 0.565 \text{ in} \qquad t_w = 14.4 \text{ mm}$$
$$t_f = 0.85 \text{ in} \qquad t_f = 21.6 \text{ mm}$$

This yields a web area A_w of

$$A_w = dt_w$$
$$= 30 \text{ in}(0.565 \text{ in}) \qquad = 762 \text{ mm}(14.4 \text{ mm})$$
$$= 16.95 \text{ in}^2 \qquad = 10{,}973 \text{ mm}^2$$

Step 5: Capacity

We now find the nominal shear capacity V_n. First, though we need to know what C_v is. We begin by finding h/t_w. We can conservatively take h as

$$h = d - 4t_f$$

$$= 30 \text{ in} - 4(0.85 \text{ in})$$

$$= 26.6 \text{ in}$$

$$= 762 \text{ mm} - 4(21.6 \text{ mm})$$

$$= 676 \text{ mm}$$

$$\frac{h}{t_w} = \frac{26.6 \text{ in}}{0.565 \text{ in}}$$

$$= 47.1$$

$$\frac{h}{t_w} = \frac{676 \text{ mm}}{14.4 \text{ mm}}$$

$$= 46.9$$

Note the small difference is due to roundoff error in the conversions. We now compare this to

$$2.24\sqrt{\frac{E}{F_y}}$$

$$2.24\sqrt{\frac{29{,}000 \text{ k/in}^2}{50 \text{ k/in}^2}}$$

$$= 53.9$$

$$2.24\sqrt{\frac{200{,}000 \text{ MN/m}^2}{345 \text{ MN/m}^2}}$$

$$= 53.9$$

Because

$$\frac{h}{t_w} \leq 2.24\sqrt{\frac{E}{F_y}},\ C_v = 1.0 \text{ and } \phi = 1.0.$$

We can now calculate nominal shear capacity.

$$V_n = 0.6F_y A_w C_v$$

$$= 0.6(50 \text{ k/in}^2)16.95 \text{ in}^2(1.0)$$

$$= 508.5 \text{ k}$$

$$= 0.6(345{,}000 \text{ kN/m}^2)0.010973 \text{ m}^2(1.0)$$

$$= 2{,}271 \text{ kN}$$

Because $\phi = 1.0$,

$$\phi V_n = 508.5 \text{ k}$$

$$\phi V_n = 2.271 \text{ kN}$$

Comparing this to the shear demand V_u, we see our beam is OK. Dividing the demand by capacity, we get a demand–capacity ratio of 0.14, meaning the web is only 14% utilized. Low DCRs for shear are common for wide flange bending members.

Step 6: Summary

Our W30 × 116 (W760 × 173) beam works.

5.7.2 Built-up Web Shear Example

Because our web works so easily, let's now look at a built-up girder. Following the previous example, we jump to Step 4, and determine the member geometry.

Step 4: Initial Size

Keeping the depth the same as the rolled section in the previous example, we will use a flange plate thickness t_f of 1.0 in (25.4 mm). Based on this, we find h, illustrated in Figure 5.13.

US	SI
$d = 30$ in	$d = 762$ mm
$h = 28.0$ in	$h = 711$ mm

We will guess a web thickness t_w of 0.25 in (6.0 mm). Remember, plate thicknesses need to be standard sizes so they are readily available.

This yields a web area A_w of

$$A_w = ht_w$$

US	SI
$= 28 \text{ in}(0.25 \text{ in})$	$= 711 \text{ mm}(6.0 \text{ mm})$
$= 7.0 \text{ in}^2$	$= 4{,}266 \text{ mm}^2$

Step 5: Capacity

Finding h/t_w

US	SI
$\frac{h}{t_w} = \frac{28 \text{ in}}{0.25 \text{ in}}$	$\frac{h}{t_w} = \frac{711 \text{ mm}}{6.0 \text{ mm}}$
$= 112$	$= 119$

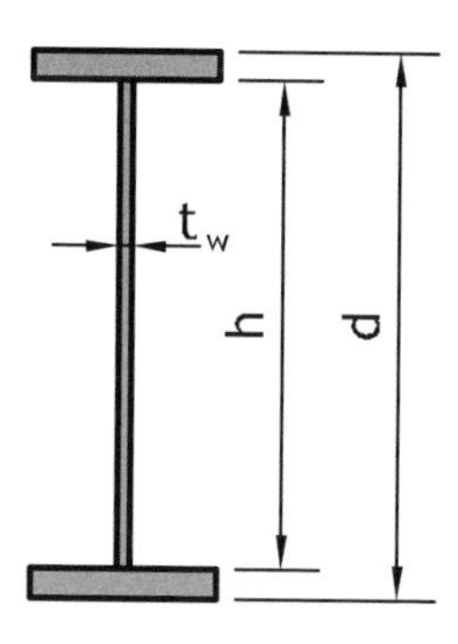

Figure 5.13 Built-up girder web shear example

Taking $k_v = 5.0$, we now compare this to

$$1.37\sqrt{\frac{k_v E}{F_y}}$$

$$1.37\sqrt{\frac{(5)29{,}000 \text{ k/in}^2}{50 \text{ k/in}^2}}$$
$$= 73.8$$

$$1.37\sqrt{\frac{(5)200{,}000 \text{ MN/m}^2}{345 \text{ MN/m}^2}}$$
$$= 73.8$$

Because $\dfrac{h}{t_w} > 1.37\sqrt{\dfrac{k_v E}{F_y}}$,

$$C_v = \frac{1.51 k_v E}{(h/t_w)^2 F_y}$$
$$= \frac{1.51(5)29{,}000 \text{ k/in}^2}{(112)^2 50 \text{ k/in}^2}$$
$$= 0.35$$

$$= \frac{1.51(5)200{,}000 \text{ MN/m}^2}{(119)^2 345 \text{ MN/m}^2}$$
$$= 0.31$$

Now to get the nominal shear capacity:

$$V_n = 0.6 F_y A_w C_v$$
$$= 0.6(50 \text{ k/in}^2)7.0 \text{ k/in}^2 (0.35)$$
$$= 73.5 \text{ k}$$

$$= 0.6(345{,}000 \text{ kN/m}^2)0.004266 \text{ m}^2 (0.31)$$
$$= 273 \text{ kN}$$

Because this is a built up shape, ϕ 0.9,

$$\phi V_n = 0.9(73.5 \text{ k})$$
$$= 66.2 \text{ k}$$

$$\phi V_n = 0.9(273 \text{ kN})$$
$$= 246 \text{ kN}$$

Unfortunately, our capacity ϕV_n is below the shear demand V_u, and our beam does not work. We can jump to the next plate size 5/16 in (8 mm) and know our web will work just fine. Perhaps you can determine the DCR for this thicker plate.

Step 6: Summary

A 5/16 in (8 mm) web plate works without stiffeners.

5.7.3 Torsion Example

Let's now look at a canopy frame that induces torsion and shear in its supporting member, shown in Figure 5.14.

Step 1: Draw Structural Layout

The canopy is supported by a HSS section on the exterior building grid and cantilevers out to cover the building entrance, shown in Figure 5.14. Key geometric values of the grid HSS are:

L_t = 12 ft
L_{cant} = 6 ft
L = 28 ft

L_t = 3.66 m
L_{cant} = 1.83 m
L = 8.54 m

Step 2: Determine Loads

Step 2a: Unit Loads

The unit loads are as follows:

$q_D = 15\ \text{lb/ft}^2$
$q_s = 60\ \text{lb/ft}^2$

$q_D = 0.718\ \text{kN/m}^2$
$q_s = 2.87\ \text{kN/m}^2$

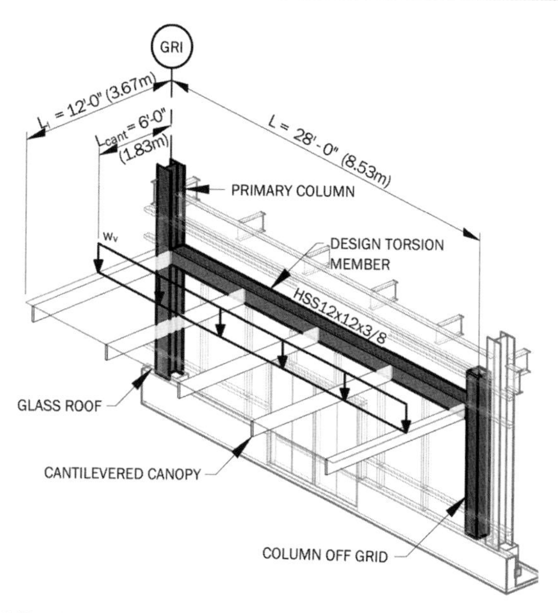

Figure 5.14 Torsion example canopy configuration

Step 2b: Load Combination

Because this is located in a region where it snows a lot, the **snow load** dominant combination will control. Multiplying the factored unit load by the canopy width L_t we find the line load w_u,

$$\begin{aligned} w_u &= [1.2q_D + 1.6q_s]L_t \\ &= \left[1.2\left(15 \text{ lb/ft}^2\right) + 1.6\left(60 \text{ lb/ft}^2\right)\right]12 \text{ ft} \\ &= 1{,}368 \text{ lb/ft} \end{aligned}$$

$$\begin{aligned} &= \left[1.2\left(0.718 \text{ kN/m}^2\right) + 1.6\left(2.87 \text{ kN/m}^2\right)\right]3.66 \text{ m} \\ &= 20.0 \text{ kN/m} \end{aligned}$$

Step 2c: Determine Member Demands

We will look at the maximum shear and torsion, which occurs at the ends.

Shear demand V_u is

$$\begin{aligned} V_u &= w_u \frac{L}{2} \\ &= 1.37 \text{ k/ft}\left(\frac{28 \text{ ft}}{2}\right) \\ &= 19.2 \text{ k} \end{aligned}$$

$$\begin{aligned} &= 20 \text{ kN/m}\left(\frac{8.54 \text{ m}}{2}\right) \\ &= 85.4 \text{ kN} \end{aligned}$$

The torsion demand T_u is similar but we add the torque arm l_{cant} into the equation

$$\begin{aligned} T_u &= w_u L_{\text{cant}} \frac{L}{2} \\ &= 1.37 \text{ k/ft}(6 \text{ ft})\left(\frac{28 \text{ ft}}{2}\right) \\ &= 115 \text{ k ft} \end{aligned}$$

$$\begin{aligned} &= 20 \text{ kN/m}(1.83 \text{ m})\left(\frac{8.54 \text{ m}}{2}\right) \\ &= 165 \text{ kN m} \end{aligned}$$

Step 3: Material Parameters

Because we are a using a rectangular HSS member, our yield stress is

$$F_y = 46 \text{ k/in}^2$$

$$F_y = 317 \text{ MN/m}^2$$

Step 4: Initial Size

The following equation gives us a quick way to get the required torsion parameter C,

$$C_{EST} = \frac{T_u}{\phi 0.45 F_y}$$

The 0.45 is an estimate to account for the shear stress reduction and direct shear in the member. We'll see how close it gets us:

$$C_{EST} = \frac{115 \text{ k ft}(12 \text{ in/ft})}{0.9(0.45)46 \text{ k/in}^2} = 74 \text{ in}^3$$

$$C_{EST} = \frac{156 \text{ kN m}}{0.9(0.45)317{,}000 \text{ kN/m}^2}\left(\frac{1{,}000 \text{ mm}}{1 \text{ m}}\right)^3 = 1.22 \times 10^6 \text{ mm}^3$$

Looking at Table A1.4, we see a HSS of 12 × 12 × 3/8 (HSS 304.8 × 304.8 × 9.5) has a larger C value:

$$C = 94.6 \text{ in}^3$$

$$C = 1.55 \times 10^6 \text{ mm}^3$$

Input geometric information is as follows:

$$d = 12.0 \text{ in} \qquad t = 0.375 \text{ in}$$

$$d = 305 \text{ mm} \qquad t = 9.5 \text{ mm}$$

$$h = d - 4t = 12 \text{ in} - 4(0.375 \text{ in}) = 10.5 \text{ in}$$

$$h = d - 4t = 305 \text{ mm} - 4(9.5 \text{ mm}) = 267 \text{ mm}$$

For direct shear, these yield a web area A_w of

$$A_w = 2ht = 2(10.5 \text{ in})0.375 \text{ in} = 7.88 \text{ in}^2$$

$$= 2(267 \text{ mm})9.5 \text{ mm} = 5{,}073 \text{ mm}^2$$

Step 5: Capacity

We now find the nominal shear and torsion capacities V_n and T_n. First, though we need to look at local web buckling. We begin by finding h/t.

$$\frac{h}{t} = \frac{10.5 \text{ in}}{0.375 \text{ in}} = 28.0$$

$$\frac{h}{t} = \frac{267 \text{ mm}}{9.5 \text{ mm}} = 28.1$$

For direct shear, we compare this to

$$1.10\sqrt{\frac{k_v E}{F_y}}$$

$$1.10\sqrt{\frac{(5)29{,}000 \text{ k/in}^2}{46 \text{ k/in}^2}} = 61.8 \qquad 1.10\sqrt{\frac{(5)200{,}000 \text{ MN/m}^2}{317 \text{ MN/m}^2}} = 61.8$$

Because $\frac{h}{t_w} \le 1.10\sqrt{\frac{k_v E}{F_y}}$, $C_v = 1.0$.

For torsion, we compare h/t to

$$2.45\sqrt{\frac{E}{F_y}}$$

$$2.45\sqrt{\frac{29{,}000 \text{ k/in}^2}{46 \text{ k/in}^2}} = 61.5 \qquad 2.45\sqrt{\frac{200{,}000 \text{ MN/m}^2}{317 \text{ MN/m}^2}} = 61.5$$

Essentially the same limit as direct shear. Which makes sense, as we are looking at local buckling and shear force in the same direction.

Almost there. We can now calculate nominal shear and torsion capacity. For shear we get,

$$V_n = 0.6F_y A_w C_v$$

$$= 0.6\left(46 \text{ k/in}^2\right)7.88 \text{ in}^2(1.0) \qquad = 0.6\left(317{,}000 \text{ kN/m}^2\right)0.005073 \text{ m}^2(1.0)$$

$$= 217.5 \text{ k} \qquad = 965 \text{ kN}$$

Applying the strength reduction factor of $\phi = 0.9$,

$$\phi V_n = 0.9(217.5 \text{ k}) \qquad \phi V_n = 0.9(965 \text{ kN})$$

$$= 195.8 \text{ k} \qquad = 868 \text{ kN}$$

Now a similar process for torsion capacity $\phi_T T_n$. The shape is compact, so we use the following equation for critical torsion stress F_{cr}

$$F_{cr} = 0.6F_y$$

$$= 0.6\left(46 \text{ k/in}^2\right) \qquad = 0.6\left(317 \text{ MN/m}^2\right)$$

$$= 27.6 \text{ k/in}^2 \qquad = 190 \text{ MN/m}^2$$

$$T_n = F_{cr}C$$

$$= 27.6 \text{ k/in}^2 \left(94.6 \text{ in}^3\right)\left(\frac{1 \text{ ft}}{12 \text{ in}}\right)$$

$$= 218 \text{ k ft}$$

$$= 190{,}000 \text{ kN/m}^2 \left(1.55 \times 10^6 \text{ mm}^3\right)\left(\frac{1 \text{ m}}{1000 \text{ mm}}\right)^3$$

$$= 295 \text{ kN m}$$

Applying the strength reduction factor of ϕ_T=0.9,

$$\phi_T T_n = 0.9(218 \text{ k ft}) = 196 \text{ k ft}$$

$$\phi_T T_n = 0.9(295 \text{ kN m}) = 266 \text{ kN m}$$

Because we are checking two failure modes that act at the same place, we use the following interaction equation:

$$\left(\frac{V_u}{\phi V_n} + \frac{T_u}{\phi T_n}\right)^2 \leq 1.0$$

Plugging in our numbers we get

$$\left(\frac{19.2 \text{ k}}{195.8 \text{ k}} + \frac{115 \text{ k ft}}{196 \text{ k ft}}\right)^2 = 0.47$$

$$\left(\frac{85.4 \text{ kN}}{868 \text{ kN}} + \frac{156 \text{ kN m}}{266 \text{ kN m}}\right)^2 = 0.47$$

Because we are less than 1.0, our combined shear and torsion member is acceptable. Circling back around to our estimate of *C*, we see we may have been unnecessarily conservative. However, we have not checked deflection, and it is possible this will control the design, as any rotation will be greatly amplified at the end of the canopy. But for now, we know the tube is strong enough.

Step 6: Summary

The HSS12 × 12 × 3/8 (HSS304.8 × 304.8 × 9.5) works for strength.

5.8 WHERE WE GO FROM HERE

This chapter provides the basis of the direct shear method and torsion for closed shapes. From here we can expand our understanding of shear in slender web elements and tension field action. This provides reductions in web members of built-up girders. We can further understand torsion in open shapes by consulting texts on advanced mechanics of materials,[2] which use the soap film analogy to find torsional shear stress.

NOTES

1. Seaburg, P.A., Carter, C.J., *Torsional Analysis of Structural Steel Members*, Steel Design Guide 9 (Chicago: American Institute of Steel Construction, 1997).
2. Cook, R.D., Young, W.C., *Advanced Mechanics of Materials* (Upper Saddle River: Prentice Hall, 1985).

Steel Compression

Chapter 6

Richard T. Seelos

Columns make open space possible and keep everything up. Without them we would be subject to the limitations of walls. Columns range from round pipe to heavy wide flanges, that are also part of the lateral force resisting system (Figure 6.1). The discussion of columns applies equally to truss compression members.

Steel, having a much larger strength and greater cost per volume than concrete, has shapes which are optimized to the point where numerous modes of failure are possible as the member receives a compressive force. In this chapter, we will discuss the failure modes, wherein the member no longer functions the way we expect.

6.1 STABILITY

6.1.1 Euler Buckling

From the analysis of materials, we know that only short columns are controlled by a yielding limit state. For taller columns, buckling is the common failure mode. Buckling is characterized by an initial deflection in the direction of load, but at a certain load, followed by much greater deflection perpendicular to the loading direction.

Figure 6.1 Heavy steel columns, Salt Lake Public Safety Building, GSBS Architects

Many people are not aware how familiar they are with buckling. If you have ever pushed along the length of a drinking straw and had it bend sideways and kink, like Figure 6.2, you have experienced global buckling.

Euler was a Swiss mathematician who studied buckling in 1759. He developed an equation for the critical load at which a column will begin to buckle (or the critical buckling load). His formulation is based on a perfectly straight, elastic, concentrically loaded member, commonly referred to as an ideal column. Most real world columns are not ideal, as they have imperfections in their geometry and material.

Euler's equation is as shown in Equation 6.1 below.

$$P_e = \frac{N^2 \pi^2 EI}{L^2} \tag{6.1}$$

where

N = number of half sine waves in the deformed shape
E = modulus of elasticity, k/in^2 (MN/m^2)
I = moment of inertia, in^4 (mm^4)
L = laterally unbraced length of member, in (mm).

While N may be any integer value (1, 2, 3 . . .), only $N = 1$ has any physical significance in the column buckling problem. Greater values of N only give larger critical load values.

Also, note that the critical buckling load is a property of the material stiffness E, and moment of inertia I (geometric stiffness), but is not dependent on the yield strength of the material.

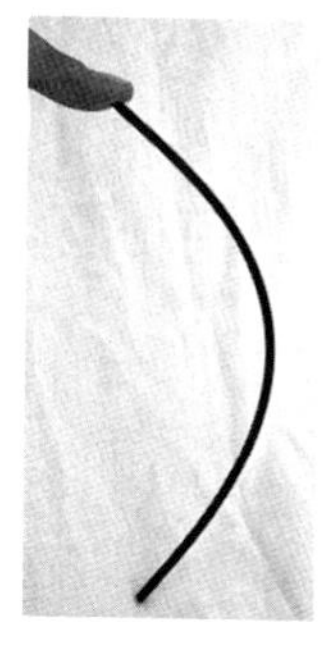

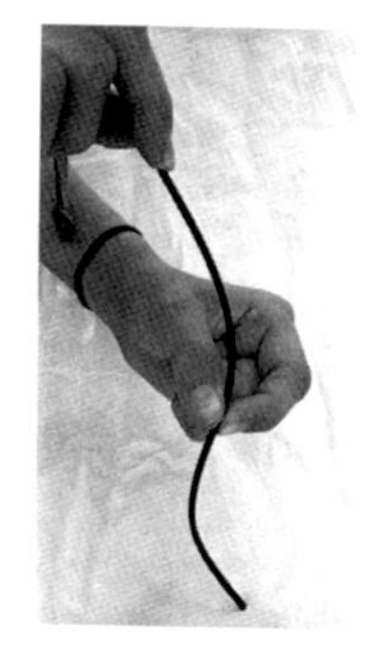

Figure 6.2 Buckling of a straw

Using the following definitions and the Euler equation we can rewrite this in terms of stress (Equation 6.2).

$$I = Ar^2$$

$$F_{cr} = \frac{P_e}{A}$$

$$P_e = \frac{N^2\pi^2 EI}{L^2} \qquad (6.2)$$

Solve for:

$$F_{cr} = \frac{P_e}{A} = \frac{N^2\pi^2 E\left(Ar^2\right)}{L^2 A}$$

$$= \frac{\pi^2 E}{\left(\frac{1}{N}\frac{L}{r}\right)^2}$$

$$F_{cr} = \frac{\pi^2 E}{\left(\frac{KL}{r}\right)^2} \qquad (6.3)$$

where

A = area, in^2 (mm^2)
F_{cr} = critical stress, k/in^2 (N/mm^2)
K = effective length factor
= (1/N)
r = radius of gyration, in (mm).

This is a more efficient method of evaluating the buckling load as it expresses the critical stress in terms of effective length factors, which are more easily observable.

6.1.2 Effective Length Factors

The effective length factor *K* is a method used to factor the member length to that of one half sine wave of the deformed shape, illustrated in Figure 6.3. We determine the effective length factor by knowing the end restraints (pinned or **fixed**) and using Figure 6.4.

For columns that are part of a frame and which do not fall easily into the **pin** or fixed conditions shown in Figure 6.4, it is still possible to determine the *K* values. We do this by using the nomographs shown in Figure 6.5. To use it we calculate the stiffness ratio Ψ at each end of the column, as follows:

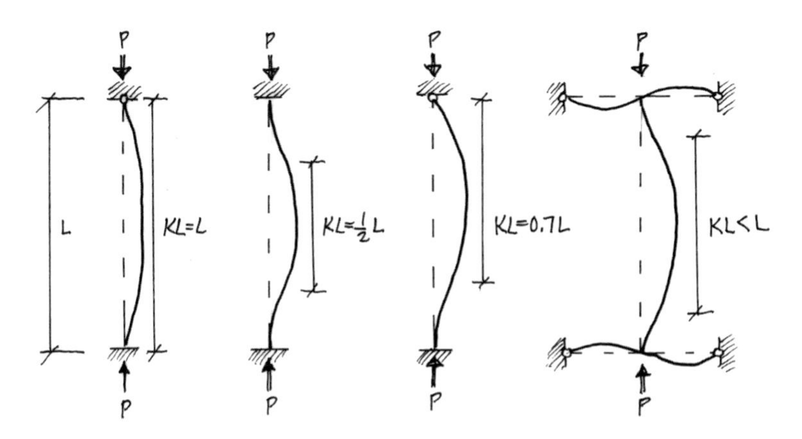

Figure 6.3 Visual depiction of effective length

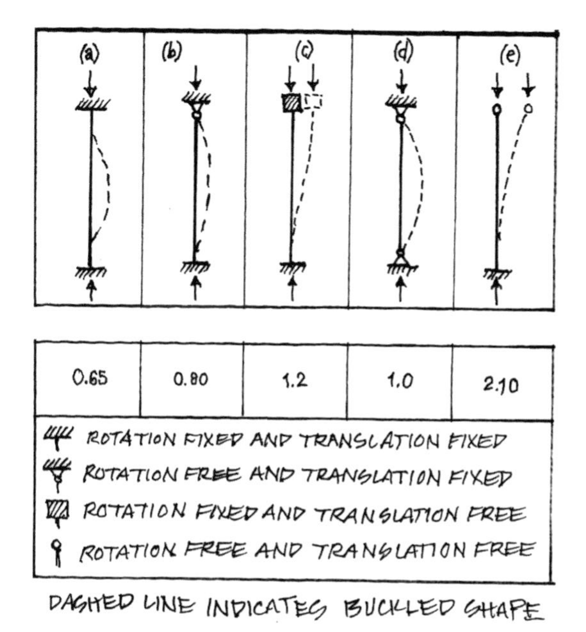

Figure 6.4 Effective length recommendations from AISC

$$\Psi = \frac{\sum(I_c/L_c)}{\sum(I_g/L_g)} \tag{6.4}$$

where

Ψ = Ψ factor

I_c = Moment of inertia of the column, in^4 (mm^4)

I_g = Moment of inertia of the girder, in^4, (mm^4)

L_c = **Effective length** of column, in (mm)

L_g = Effective length of girder in, (mm).

Drawing a line in Figure 6.5 from the calculated Ψ_A to the calculated Ψ_B will give the effective length factor where the resulting line crosses the scale in the middle.

There are two charts: non-sway and sway. Non-sway frames have little deflection under lateral loads, and are consistent with braced frames and shear walls. **Sway frames** have more lateral deflection, and are commonly **moment frames**. On both charts, Ψ may be taken as 10.0 for pin connections and 1.0 for fixed connections.

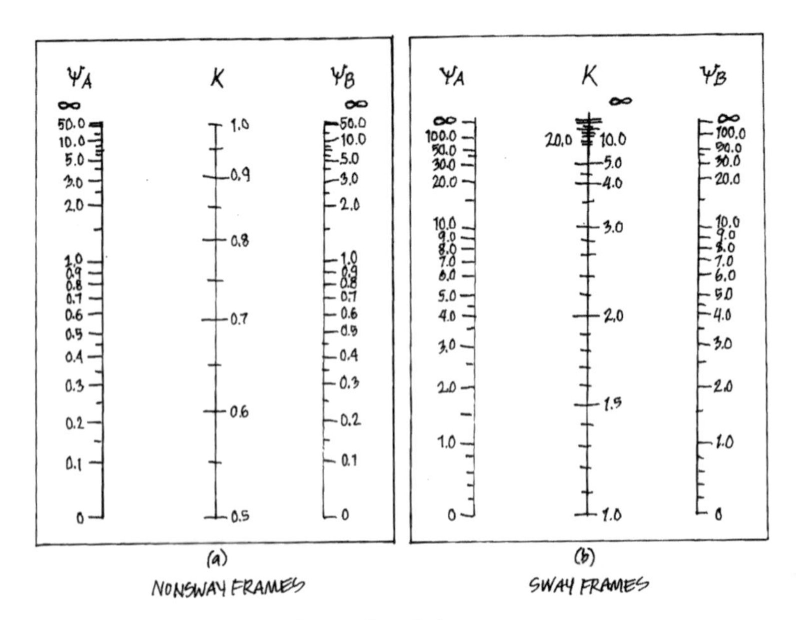

Figure 6.5 Alignment charts for members in frames

6.1.3 Slenderness

Now that we understand the format of the effective length factors, it is possible to understand the slenderness requirement for columns. For compression members, the code recommends limiting the column slenderness according to Equation 6.5.

$$\frac{L_c}{r} \le 200 \tag{6.5}$$

where

L_c = effective length of member, in (mm)
 = KL
K = Effective length factor
L = Laterally unbraced length of member, in (mm)
r = Least radius of gyration, in (mm).

By following this recommendation, we are comfortable that the imperfections in our columns will not be significant enough to accidentally push our column into a global buckling instability failure mode prior to reaching our expected capacity. Additionally, when we get outside of this range, the critical buckling stress is so low that we will likely have to use a larger column anyway.

6.2 CAPACITY

When it comes to compression loads, there are three types of failure; yielding, global buckling, and local buckling. We will discuss yielding and global buckling below. See Chapter 2 for local buckling.

6.2.1 Yielding

Yielding occurs when stresses in the member get large enough to cause the material to extend into the plastic range (Figure 2.4). After yielding in compression, the height decreases, illustrated in Figure 6.6, and the volume remains constant. This causes the cross-section to increase as shown in Figure 6.7. Yielding controls short column failure. In truth, we do not have a separate check for yielding but limit buckling failure modes by the yield capacity.

6.1.2 Global Buckling

Global buckling is where the entire member laterally deforms over the length of the member, like the drinking straw. In plan view, the member translates relative to its original position, illustrated in Figure 6.8. The deformation is greatest at the middle of the column.

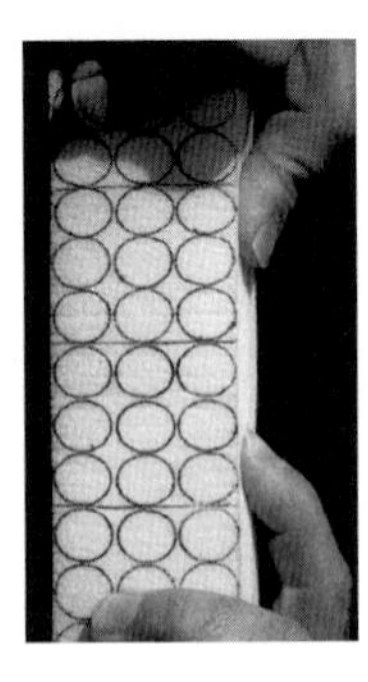

Figure 6.6 Compression deformed shape in foam column

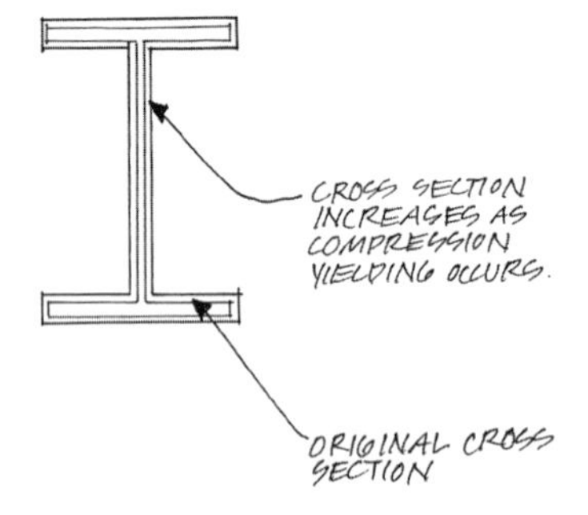

Figure 6.7 Deformed shape due to yielding

The story of Samson in the Bible[1] is a good example of global buckling. At the end of his story, Samson with eyes put out, was brought to the temple and set between two pillars. Samson then pushed on the pillars and brought the whole building down. Whether true or not, the two building columns must have been close to their global buckling capacity and just needed a little push to instigate global buckling.

Column strength is a function of yielding, and the two modes of global buckling; inelastic and elastic, are shown in Figure 6.9. When L_c/r is small yielding controls strength. Next, inelastic buckling occurs, where the deformation is a mix of yielding and elastic stretching. The last region is elastic buckling.

We would expect that the strength of a column would be limited by the yielding strength (shown by the location of where the capacity line crosses

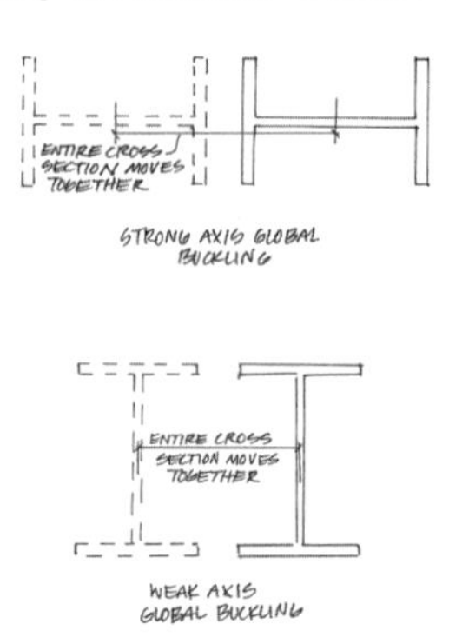

Figure 6.8 Deformed shape due to global buckling

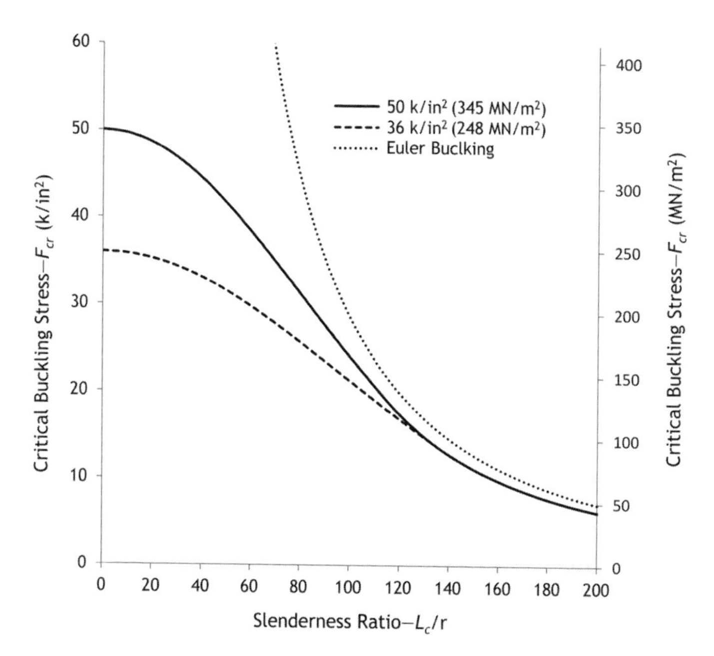

Figure 6.9 Buckling stress as a function of slenderness

the y axis) until it crossed with the Euler buckling limitations (shown by the curved dashed line). In reality, testing shows the critical stress falls below the yield strength long before it approaches the Euler buckling limitation.

Nominal column capacity, P_n, is given by Equation 6.6:

$$P_n = F_{cr} A_g \tag{6.6}$$

where

F_{cr} = critical buckling stress, k/in² (MN/m²), given below
A_g = gross area, in² (mm²)
ϕ = 0.9 for compression limit states.

Unlike tension elements, we are not concerned with net area but only use the gross area. This is because the material is being compressed and won't rupture through holes, and buckling is a function of the stiffness over the length, not a local point.

The two equations below approximate the buckling curve observed by testing, shown in Figure 6.9. The point that defines the transition between the equations is defined by relationship of the slenderness ratio L_c/r to:

$$4.71\sqrt{\frac{E}{F_y}} \tag{6.7}$$

where

F_y = Yield stress, k/in² (N/mm²)
E = Modulus of elasticity, in⁴ (mm⁴).

When L_c/r is less than this value, the column is controlled by inelastic buckling. When greater, elastic buckling dominates.

When $\frac{L_c}{r} \leq 4.71\sqrt{\frac{E}{F_y}}$,

we use the inelastic equation for critical buckling stress, given by equation Figure 6.8. This means that if the column were to fail, it would experience some yielding before it finally buckled. If the load was then removed, before it buckled completely, the column would not return to the same position before the buckling occurred.

$$F_{cr} = \left(0.658^{F_y/F_e}\right)F_y \tag{6.8}$$

where

F_y = yield stress, k/in² (N/mm²)
F_e = Elastic buckling stress, k/in² (N/mm²), given by Equation 6.9:

$$F_e = \frac{\pi^2 E}{(L_c/r)^2} \tag{6.9}$$

where

E = Modulus of elasticity, in⁴ (mm⁴).

When $\frac{L_c}{r} > 4.71\sqrt{\frac{E}{F_y}}$,

the column buckles in the elastic range, without experiencing yielding. If the load was removed, the member would snap back to its original geometry. The critical buckling stress in this range is given by Equation 6.10:

$$F_{cr} = 0.877F_e \tag{6.10}$$

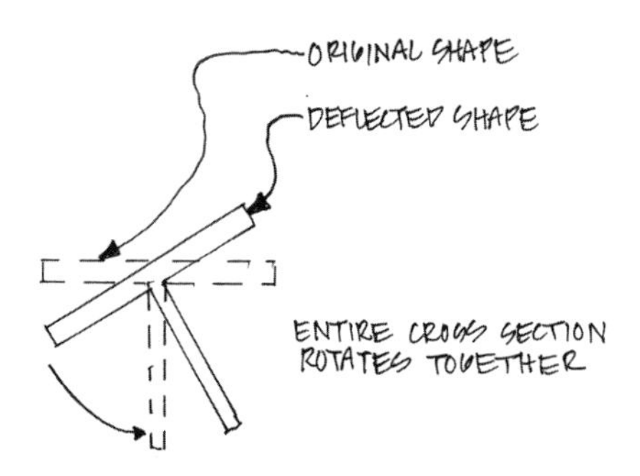

Figure 6.10 Shape deformation due to torsional buckling

This is essentially the Euler equation we previously derived, multiplied by 0.877 to account for possible lack of straightness and loading eccentricity. Remember to multiply the nominal capacity P_n by ϕ to get design strength ϕP_n.

6.2.3 Torsional Buckling

Torsional Buckling occurs only in members made up entirely of unstiffened elements. It is characterized by twisting of the cross section, shown in Figure 6.10. This mode may occur with WT sections and double angles, but would not occur with W-sections or channels. We generally avoid using these types of sections as compression elements.

Initial Column Sizing

To find a good starting point for column size, Table 6.1 provides the expected column sizes based on the number of stories or tributary area.

Assuming we know the column load, the next step in initial sizing is estimating the slenderness L_c/r and determining the critical buckling stress from Table 6.2. Selecting L_c/r from 100 to 150 is a good starting point.

We then divide the column load by critical buckling stress to get the estimated column area A_{est}, shown in Equation 6.11:

$$A_{est} = \frac{P_u}{\phi F_{cr}} \tag{6.11}$$

Table 6.1 Estimated typical column sizes

Tributary Area (ft^2)	*Imperial Shapes*		Number Stories for Approximately 25'–0 (7.5 m) Square Bay	*Tributary Area (m^2)*	*Metric Shapes*	
250	W4 × 13	HSS4 × 4 × 3/16		25	W100 × 19.3	HSS101.6x101.6x4.8
500	W6 × 16	HSS4 × 4 × 1/4	1	35	W150 × 24.0	HSS101.6 × 101.6 × 6.4
1,000	W6 × 20	HSS4 × 4 × 3/8	2	90	W150 × 29.8	HSS101.6 × 101.6 × 9.5
	W8 × 21	HSS5 × 5 × 1/4			W200 × 31.3	HSS127 × 127 × 6.4
1,500	W8 × 24	HSS5 × 5 × 3/8	3	140	W200 × 35.9	HSS127 × 127 × 9.5
	W10 × 30	HSS6 × 6 × 1/4			W250 × 44.8	HSS152.4 × 152.4 × 6.4
2,000	W8 × 31	HSS5 × 5 × 1/2	4	190	W200 × 46.1	HSS127 × 127 × 12.7
	W10 × 33	HSS6 × 6 × 3/8			W250 × 49.1	HSS152.4 × 152.4 × 9.5
3,000	W8 × 40	HSS6 × 6 × 5/8	5	280	W200 × 59	HSS152.4 × 152.4 × 15.9
	W10 × 39	HSS8 × 8 × 3/8			W250 × 58	HSS203.2 × 203.2 × 9.5
4,000	W8 × 67	HSS8 × 8 × 1/2	7	370	W200 × 100	HSS203.2 × 203.2 × 12.7
	W10 × 49				W250 × 73	
5,000	W10 × 54	HSS8 × 8 × 5/8	8	470	W250 × 80	HSS203.2 × 203.2 × 15.9
	W12 × 58	HSS10 × 10 × 1/2			W310 × 86	HSS254 × 254 × 12.7
6,000	W10 × 77	HSS10 × 10 × 5/8	10	560	W250 × 115	HSS254 × 254 × 15.9
	W12 × 72				W310 × 107	

8,000	W10 × 100			13	750	W250 × 149	
	W12 × 87	HSS12 × 12 × 1/2				W310 × 129	HSS304.8 × 304.8 × 12.7
10,000	W10 × 112	HSS12 × 12 × 5/8		16	900	W250 × 167	HSS304.8 × 304.8 × 15.9
	W12 × 106	HSS14 × 14 × 3/8				W310 × 158	HSS355.6 × 355.6 × 15.9
	W14 × 99					W360 × 147	
20,000	W12 × 230	W14x257		32	1800	W310 × 342	W360 × 382
40,000	W14 × 398			65	3700	W360 × 592	
60,000	W14 × 550			100	5600	W360 × 818	
90,000	W14 × 730			150	8500	W360 × 1086	

Notes

1) This table is for preliminary sizing only. Final section sizes must be calculated based on actual loading, length, and section size

2) For normal height columns (10–12') and moderate loading

3) For heavy loads, increase tributary area by 15%; for light loads, decrease area by 10%

4) To extend table to other shapes match A, I_y, r_y

Table 6.2 Critical buckling stress based on L_c/r

	ϕF_{cr} (k/in^2)																		
L/r	F_y (k/in^2)			L/r	F_y (k/in^2)			L/r	F_y (k/in^2)			L/r	F_y (k/in^2)			L/r	F_y (k/in^2)		
	36	46	50		36	46	50		36	46	50		36	46	50		36	46	50
2	32.4	41.4	45.0	42	29.5	36.8	39.5	82	22.7	26.3	27.5	122	14.8	15.2	15.2	162	8.61	8.61	8.61
4	32.4	41.4	44.9	44	29.3	36.3	39.1	84	22.3	25.8	26.9	124	14.4	14.7	14.7	164	8.40	8.40	8.40
6	32.3	41.3	44.9	46	29.0	35.9	38.5	86	22.0	25.2	26.2	126	14.0	14.2	14.2	166	8.20	8.20	8.20
8	32.3	41.2	44.8	48	28.7	35.4	38.0	88	21.6	24.6	25.5	128	13.7	13.8	13.8	168	8.00	8.00	8.00
10	32.2	41.1	44.7	50	28.4	35.0	37.5	90	21.2	24.0	24.9	130	13.3	13.4	13.4	170	7.82	7.82	7.82
12	32.2	41.0	44.5	52	28.1	34.5	36.9	92	20.8	23.4	24.2	132	12.9	13.0	13.0	172	7.64	7.64	7.64
14	32.1	40.9	44.4	54	27.8	34.0	36.4	94	20.3	22.8	23.6	134	12.6	12.6	12.6	174	7.46	7.46	7.46
16	32.0	40.7	44.2	56	27.5	33.5	35.8	96	19.9	22.3	22.9	136	12.2	12.2	12.2	176	7.29	7.29	7.29
18	31.9	40.5	43.9	58	27.1	33.0	35.2	98	19.5	21.7	22.3	138	11.9	11.9	11.9	178	7.13	7.13	7.13
20	31.7	40.3	43.7	60	26.8	32.5	34.6	100	19.1	21.1	21.7	140	11.5	11.5	11.5	180	6.97	6.97	6.97
22	31.6	40.1	43.4	62	26.5	32.0	34.0	102	18.7	20.6	21.0	142	11.2	11.2	11.2	182	6.82	6.82	6.82
24	31.4	39.8	43.1	64	26.1	31.4	33.4	104	18.3	20.0	20.4	144	10.9	10.9	10.9	184	6.67	6.67	6.67
26	31.3	39.6	42.8	66	25.8	30.9	32.7	106	17.9	19.4	19.8	146	10.6	10.6	10.6	186	6.53	6.53	6.53
28	31.1	39.3	42.5	68	25.4	30.3	32.1	108	17.5	18.9	19.2	148	10.3	10.3	10.3	188	6.39	6.39	6.39

30	30.9	39.0	42.1	70	25.0	29.8	31.4	110	17.1	18.3	18.6	150	10.0	10.0	10.0	190	6.26	6.26	6.26
32	30.7	38.6	41.8	72	24.7	29.2	30.8	112	16.7	17.8	18.0	152	9.78	9.78	9.78	192	6.13	6.13	6.13
34	30.5	38.3	41.4	74	24.3	28.6	30.2	114	16.3	17.3	17.4	154	9.53	9.53	9.53	194	6.00	6.00	6.00
36	30.3	37.9	40.9	76	23.9	28.1	29.5	116	16.0	16.7	16.8	156	9.28	9.28	9.28	196	5.88	5.88	5.88
38	30.0	37.6	40.5	78	23.5	27.5	28.8	118	15.6	16.2	16.2	158	9.05	9.05	9.05	198	5.76	5.76	5.76
40	29.8	37.2	40.0	80	23.1	26.9	28.2	120	15.2	15.7	15.7	160	8.82	8.82	8.82	200	5.65	5.65	5.65

Source: AISC Steel Construction Manual, 14th Edition

Table 6.2m Critical buckling stress based on L_c/r

	ϕF_{cr} (MN/m^2)																		
L/r	F_y (MN/m^2)			L/r	F_y (MN/m^2)			L/r	F_y (MN/m^2)			L/r	F_y (MN/m^2)			L/r	F_y (MN/m^2)		
	250	320	345		250	320	345		250	320	345		250	320	345		250	320	345
2	223	285	310	42	203	254	272	82	157	181	190	122	102	105	105	162	59.4	59.4	59.4
4	223	285	310	44	202	250	270	84	154	178	185	124	99.3	101	101	164	57.9	57.9	57.9
6	223	285	310	46	200	248	265	86	152	174	181	126	96.5	97.9	97.9	166	56.5	56.5	56.5
8	223	284	309	48	198	244	262	88	149	170	176	128	94.5	95.1	95.1	168	55.2	55.2	55.2
10	222	283	308	50	196	241	259	90	146	165	172	130	91.7	92.4	92.4	170	53.9	53.9	53.9
12	222	283	307	52	194	238	254	92	143	161	167	132	88.9	89.6	89.6	172	52.7	52.7	52.7
14	221	282	306	54	192	234	251	94	140	157	163	134	86.9	86.9	86.9	174	51.4	51.4	51.4
16	221	281	305	56	190	231	247	96	137	154	158	136	84.1	84.1	84.1	176	50.3	50.3	50.3
18	220	279	303	58	187	228	243	98	134	150	154	138	82.0	82.0	82.0	178	49.2	49.2	49.2
20	219	278	301	60	185	224	239	100	132	145	150	140	79.3	79.3	79.3	180	48.1	48.1	48.1
22	218	276	299	62	183	221	234	102	129	142	145	142	77.2	77.2	77.2	182	47.0	47.0	47.0
24	216	274	297	64	180	216	230	104	126	138	141	144	75.2	75.2	75.2	184	46.0	46.0	46.0
26	216	273	295	66	178	213	225	106	123	134	137	146	73.1	73.1	73.1	186	45.0	45.0	45.0
28	214	271	293	68	175	209	221	108	121	130	132	148	71.0	71.0	71.0	188	44.1	44.1	44.1

30	213	269	290	70	172	205	216	110	118	126	128	150	68.9	68.9	68.9	190	43.2	43.2	43.2
32	212	266	288	72	170	201	212	112	115	123	124	152	67.4	67.4	67.4	192	42.3	42.3	42.3
34	210	264	285	74	168	197	208	114	112	119	120	154	65.7	65.7	65.7	194	41.4	41.4	41.4
36	209	261	282	76	165	194	203	116	110	115	116	156	64.0	64.0	64.0	196	40.5	40.5	40.5
38	207	259	279	78	162	190	199	118	108	112	112	158	62.4	62.4	62.4	198	39.7	39.7	39.7
40	205	256	276	80	159	185	194	120	105	108	108	160	60.8	60.8	60.8	200	39.0	39.0	39.0

Source: AISC Steel Construction Manual, 14th Edition

Table 6.3 Wide flange axial compressive strength

	Axial Compressive Strength, $\phi_c P_n$ (k)											
L_C	*W14x*			*W12x*					*W10x*			*W8x*
(ft)	*730*	*257*	*90*	*336*	*230*	*152*	*96*	*58*	*112*	*49*	*33*	*67*
0	9,670	3,400	1,190	4,450	3,050	2,010	1,270	765	1,480	648	437	886
6	9,510	3,330	1,160	4,310	2,940	1,940	1,220	720	1,400	611	395	815
7	9,450	3,300	1,150	4,260	2,910	1,910	1,200	705	1,380	598	381	790
8	9,380	3,270	1,140	4,210	2,860	1,880	1,180	687	1,350	584	365	763
9	9,310	3,240	1,120	4,150	2,820	1,850	1,160	668	1,310	568	348	733
10	9,220	3,200	1,100	4,080	2,770	1,810	1,140	647	1,280	550	330	701
11	9,130	3,160	1,090	4,000	2,710	1,770	1,110	625	1,240	532	311	668
12	9,030	3,110	1,070	3,920	2,650	1,730	1,080	601	1,200	512	292	633
13	8,920	3,060	1,050	3,840	2,590	1,690	1,050	577	1,160	492	272	597
14	8,810	3,010	1,030	3,750	2,520	1,640	1,020	551	1,110	471	253	560
15	8,690	2,960	1,000	3,660	2,450	1,590	990	525	1,060	449	233	523
16	8,560	2,900	979	3,560	2,380	1,540	957	499	1,020	427	214	487
18	8,290	2,790	929	3,350	2,230	1,440	888	445	921	382	177	415
20	7,990	2,660	877	3,140	2,070	1,330	816	392	824	337	143	347

24	7,340	2,380	766	2,690	1,750	1,110	672	292	636	253	99.5	241
28	6,650	2,100	653	2,240	1,430	894	535	214	473	186	73.1	177
32	5,930	1,810	543	1,820	1,140	697	413	164	362	142	56	136
36	5,200	1,530	439	1,440	898	551	326	130	286	112		
40	4,500	1,260	356	1,170	727	446	264	105	232	91.1		
44	3,830											
48	3,220											
50	2,970											

Source: AISC Steel Construction Manual, 14th Edition

Table 6.3m **Wide flange axial compressive strength**

	Axial Compressive Strength, $\phi_c P_n$ (kN)											
L_C	*W360*			*W310*					*W250*			*W200*
(m)	*1086*	*382*	*134*	*500*	*342*	*226*	*143*	*86*	*167*	*73*	*49.1*	*100*
0	43,014	15,124	5,293	19,795	13,567	8,941	5,649	3,403	6,583	2,882	1,944	3,941
1.8	42,303	14,813	5,160	19,172	13,078	8,630	5,427	3,203	6,228	2,718	1,757	3,625
2.1	42,036	14,679	5,115	18,949	12,944	8,496	5,338	3,136	6,139	2,660	1,695	3,514
2.4	41,724	14,546	5,071	18,727	12,722	8,363	5,249	3,056	6,005	2,598	1,624	3,394
2.7	41,413	14,412	4,982	18,460	12,544	8,229	5,160	2,971	5,827	2,527	1,548	3,261
3.0	41,013	14,234	4,893	18,149	12,322	8,051	5,071	2,878	5,694	2,447	1,468	3,118
3.4	40,612	14,056	4,849	17,793	12,055	7,873	4,938	2,780	5,516	2,366	1,383	2,971
3.7	40,167	13,834	4,760	17,437	11,788	7,695	4,804	2,673	5,338	2,277	1,299	2,816
4.0	39,678	13,612	4,671	17,081	11,521	7,517	4,671	2,567	5,160	2,189	1,210	2,656
4.3	39,189	13,389	4,582	16,681	11,210	7,295	4,537	2,451	4,938	2,095	1,125	2,491
4.6	38,655	13,167	4,448	16,280	10,898	7,073	4,404	2,335	4,715	1,997	1,036	2,326
4.9	38,077	12,900	4,355	15,836	10,587	6,850	4,257	2,220	4,537	1,899	952	2,166
5.5	36,876	12,411	4,132	14,902	9,920	6,405	3,950	1,979	4,097	1,699	787	1,846
6.1	35,541	11,832	3,901	13,967	9,208	5,916	3,630	1,744	3,665	1,499	636	1,544

7.3	32,650	10,587	3,407	11,966	7,784	4,938	2,989	1,299	2,829	1,125	443	1,072
8.5	29,581	9,341	2,905	9,964	6,361	3,977	2,380	952	2,104	827	325	787
9.8	26,378	8,051	2,415	8,096	5,071	3,100	1,837	730	1,610	632	249	605
11.0	23,131	6,806	1,953	6,405	3,995	2,451	1,450	578	1,272	498		
12.2	20,017	5,605	1,584	5,204	3,234	1,984	1,174	467	1,032	405		
13.4	17,037											
14.6	14,323											
15.2	13,211											

Source: AISC Steel Construction Manual, 14th Edition

Table 6.4 Square HSS axial compressive strength

	Axial Compressive Strength, $\phi_c P_n$ (k)																
L_C	HSS16x16x		HSS14x14x		HSS12x12x		HSS10x10x		HSS8x8x		HSS6x6x			HSS4x4x		HSS 3x3	
(ft)	1/2	3/8	5/8	3/8	5/8	3/8	5/8	3/8	5/8	3/8	5/8	3/8	1/4	3/8	1/4	1/4	1/8
0	1,170	782	1,250	748	1,060	662	869	546	679	431	484	314	217	198	140	101	53.8
4	1,170	779	1,250	743	1,060	660	860	541	667	424	467	305	211	184	130	89.1	48.1
6	1,160	779	1,240	743	1,050	652	849	534	653	415	450	293	204	168	120	76.1	41.7
7	1,160	777	1,230	741	1,040	648	841	530	644	410	438	286	199	159	114	68.7	38.1
8	1,150	776	1,230	738	1,030	644	833	525	633	404	425	279	194	149	107	61.1	34.2
9	1,150	774	1,220	736	1,030	640	823	519	622	397	410	270	188	138	99.3	53.4	30.3
10	1,140	772	1,210	733	1,020	634	813	513	609	389	394	260	182	126	91.7	46	26.5
11	1,140	769	1,210	729	1,010	629	802	506	596	381	378	250	175	115	84	39	22.9
12	1,130	767	1,200	726	997	622	789	499	581	372	360	240	168	104	76.3	32.8	19.4
13	1,120	764	1,190	722	985	616	776	491	565	363	342	229	161	92.8	68.7	27.9	16.5
14	1,120	761	1,180	718	973	609	762	483	549	353	324	218	153	82.2	61.3	24.1	14.2
15	1,110	758	1,170	713	961	601	748	474	532	343	305	206	146	72	54.3	21	12.4
16	1,100	755	1,150	708	947	593	732	465	514	333	286	195	138	63.3	47.7	18.4	10.9
18	1,080	747	1,130	697	918	576	700	446	478	311	249	172	122	50	37.7	14.6	8.62

20	1,060	739	1,100	683	887	557	665	425	440	288	213	149	107	40.5	30.5		
22	1,040	730	1,070	665	854	537	628	403	402	264	179	127	92.1	33.5	25.2		
24	1,020	720	1,040	646	819	516	591	380	364	241	150	107	78.1	28.1	21.2		
26	994	709	1,010	626	783	494	552	357	326	218	128	91.4	66.6				
28	968	697	970	605	745	472	514	333	290	195	110	78.8	57.4				
30	941	684	934	583	707	449	475	310	256	174	96	68.7	50				
34	884	656	859	538	630	402	400	264	199	136	74.8	53.5	38.9				
38	825	623	782	492	552	354	329	220	159	109		42.8	31.2				
40	794	606	743	468	515	331	297	199	144	98.0							

Source: AISC Steel Construction Manual, 14th Edition

Table 6.4m Square HSS axial compressive strength

	Axial Compressive Strength, $\phi_c P_n$ (kN)																
L_C	*HSS406.4x406.4*		*HSS355.6x355.6*		*HSS304.8x304.8*		*HSS254x254*		*HSS203.2x203.2*		*HSS152.4x152.4*			*HSS101.6x101.6*		*HSS76.2x76.2*	
(m)	*15.9*	*9.5*	*15.9*	*9.5*	*15.9*	*9.5*	*15.9*	*9.5*	*15.9*	*9.5*	*15.9*	*9.5*	*6.4*	*9.5*	*6.4*	*6.4*	*3.2*
0	5,204	3,479	5,560	3,327	4,715	2,945	3,866	2,429	3,020	1,917	2,153	1,397	965	881	623	449	239
1.2	5,204	3,465	5,560	3,305	4,715	2,936	3,825	2,406	2,967	1,886	2,077	1,357	939	818	578	396	214
1.8	5,160	3,465	5,516	3,305	4,671	2,900	3,777	2,375	2,905	1,846	2,002	1,303	907	747	534	339	185
2.1	5,160	3,456	5,471	3,296	4,626	2,882	3,741	2,358	2,865	1,824	1,948	1,272	885	707	507	306	169
2.4	5,115	3,452	5,471	3,283	4,582	2,865	3,705	2,335	2,816	1,797	1,890	1,241	863	663	476	272	152
2.7	5,115	3,443	5,427	3,274	4,582	2,847	3,661	2,309	2,767	1,766	1,824	1,201	836	614	442	238	135
3.0	5,071	3,434	5,382	3,261	4,537	2,820	3,616	2,282	2,709	1,730	1,753	1,157	810	560	408	205	118
3.4	5,071	3,421	5,382	3,243	4,493	2,798	3,567	2,251	2,651	1,695	1,681	1,112	778	512	374	173	102
3.7	5,026	3,412	5,338	3,229	4,435	2,767	3,510	2,220	2,584	1,655	1,601	1,068	747	463	339	146	86.3
4.0	4,982	3,398	5,293	3,212	4,381	2,740	3,452	2,184	2,513	1,615	1,521	1,019	716	413	306	124	73.4
4.3	4,982	3,385	5,249	3,194	4,328	2,709	3,390	2,148	2,442	1,570	1,441	970	681	366	273	107	63.2
4.6	4,938	3,372	5,204	3,172	4,275	2,673	3,327	2,108	2,366	1,526	1,357	916	649	320	242	93.4	55.2
4.9	4,893	3,358	5,115	3,149	4,212	2,638	3,256	2,068	2,286	1,481	1,272	867	614	282	212	81.8	48.5
5.5	4,804	3,323	5,026	3,100	4,083	2,562	3,114	1,984	2,126	1,383	1,108	765	543	222	168	64.9	38.3

6.1	4,715	3,287	4,893	3,038	3,946	2,478	2,958	1,890	1,957	1,281	947	663	476	180	136			
6.7	4,626	3,247	4,760	2,958	3,799	2,389	2,793	1,793	1,788	1,174	796	565	410	149	112			
7.3	4,537	3,203	4,626	2,874	3,643	2,295	2,629	1,690	1,619	1,072	667	476	347	125	94.3			
7.9	4,422	3,154	4,493	2,785	3,483	2,197	2,455	1,588	1,450	970	569	407	296					
8.5	4,306	3,100	4,315	2,691	3,314	2,100	2,286	1,481	1,290	867	489	351	255					
9.1	4,186	3,043	4,155	2,593	3,145	1,997	2,113	1,379	1,139	774	427	306	222					
10.4	3,932	2,918	3,821	2,393	2,802	1,788	1,779	1,174	885	605	333	238	173					
11.6	3,670	2,771	3,479	2,189	2,455	1,575	1,463	979	707	485		190	139					
12.2	3,532	2,696	3,305	2,082	2,291	1,472	1,321	885	641	436								

Source: AISC Steel Construction Manual, 14th Edition

Table 6.5 Round pipe axial compressive strength

	Axial Compressive Strength, $\phi_c P_n$ (k)													
L_C	*Pipe 12*		*Pipe 10*		*Pipe 8*		*Pipe 6*		*Pipe 5*		*Pipe 4*		*Pipe 3*	
(ft)	*XS*	*STD*	*XS*	*STD*	*XS*	*STD*	*XS*	*STD*	*XS*	*STD*	*XS*	*STD*	*XS*	*STD*
0	551	432	476	362	375	247	247	164	180	126	130	93.2	89.1	65.2
6	544	426	466	355	363	240	233	155	167	117	116	83	72.7	53.7
7	541	424	463	353	359	237	229	153	162	114	111	79.6	67.5	50.1
8	538	421	459	350	354	234	224	149	157	111	105	75.8	62.0	46.2
9	534	418	455	347	349	231	218	146	152	107	99.3	71.8	56.3	42.2
10	530	415	450	343	343	227	212	142	146	103	93.1	67.5	50.6	38.1
11	526	412	445	339	337	223	205	137	139	98.1	86.8	63.1	44.9	34.0
12	521	408	439	335	330	219	198	133	132	93.6	80.3	58.5	39.4	30.0
13	516	405	433	330	323	214	191	128	125	88.8	73.8	54	34.1	26.2
14	511	400	427	326	315	209	183	123	118	83.9	67.4	49.5	29.4	22.7
15	505	396	420	320	307	204	175	118	111	79.0	61.2	45.1	25.6	19.8
16	499	391	413	315	299	199	167	113	104	74.1	55.1	40.8	22.5	17.4
18	486	381	397	304	282	188	151	102	89.8	64.3	43.9	32.7	17.8	13.7
20	472	370	381	291	263	176	134	91.5	76.3	54.9	35.6	26.5		

	22	457	359	363	278	245	164	118	81	63.6	45.9	29.4	21.9	
	24	440	346	345	265	225	152	103	70.8	53.4	38.6	24.7	18.4	
	26	424	333	327	251	206	139	88.0	61.1	45.5	32.9			
	28	406	320	308	236	188	127	75.8	52.7	39.2	28.4			
	30	388	306	288	222	169	115	66.1	45.9	34.2	24.7			
	34	351	277	250	193	135	92.7	51.4	35.7					
	38	314	248	213	165	108	74.2							
	40	296	234	195	152	97.5	67							

Source: AISC Steel Construction Manual, 14th Edition

Table 6.5m Round pipe axial compressive strength

	Axial Compressive Strength, $\phi_c P_n$ (kN)													
L_C	*Pipe 310*		*Pipe 254*		*Pipe 203*		*Pipe 152*		*Pipe 127*		*Pipe 102*		*Pipe 75*	
(m)	*XS*	*STD*	*XS*	*STD*	*XS*	*STD*	*XS*	*STD*	*XS*	*STD*	*XS*	*STD*	*XS*	*STD*
0	2,451	1,922	2,117	1,610	1,668	1,099	1,099	730	801	560	578	415	396	290
1.8	2,420	1,895	2,073	1,579	1,615	1,068	1,036	689	743	520	516	369	323	239
2.1	2,406	1,886	2,060	1,570	1,597	1,054	1,019	681	721	507	494	354	300	223
2.4	2,393	1,873	2,042	1,557	1,575	1,041	996	663	698	494	467	337	276	206
2.7	2,375	1,859	2,024	1,544	1,552	1,028	970	649	676	476	442	319	250	188
3.0	2,358	1,846	2,002	1,526	1,526	1,010	943	632	649	458	414	300	225	169
3.4	2,340	1,833	1,979	1,508	1,499	992	912	609	618	436	386	281	200	151
3.7	2,318	1,815	1,953	1,490	1,468	974	881	592	587	416	357	260	175	133
4.0	2,295	1,802	1,926	1,468	1,437	952	850	569	556	395	328	240	152	117
4.3	2,273	1,779	1,899	1,450	1,401	930	814	547	525	373	300	220	131	101
4.6	2,246	1,761	1,868	1,423	1,366	907	778	525	494	351	272	201	114	88.1
4.9	2,220	1,739	1,837	1,401	1,330	885	743	503	463	330	245	181	100	77.4
5.5	2,162	1,695	1,766	1,352	1,254	836	672	454	399	286	195	145	79.2	60.9
6.1	2,100	1,646	1,695	1,294	1,170	783	596	407	339	244	158	118		

6.7	2,033	1,597	1,615	1,237	1,090	730	525	360	283	204	131	97.4		
7.3	1,957	1,539	1,535	1,179	1,001	676	458	315	238	172	110	81.8		
7.9	1,886	1,481	1,455	1,117	916	618	391	272	202	146				
8.5	1,806	1,423	1,370	1,050	836	565	337	234	174	126				
9.1	1,726	1,361	1,281	988	752	512	294	204	152	110				
10.4	1,561	1,232	1,112	859	601	412	229	159						
11.6	1,397	1,103	947	734	480	330								
12.2	1,317	1,041	867	676	434	298								

Source: AISC Steel Construction Manual, 14th Edition

where

P_u = axial demand, k (kN)
ϕF_{cr} = critical buckling stress, k/in^2 (MN/m^2)
ϕ_r = strength reduction factor, 0.9 for compression.

This area can then be used to quickly choose a shape which can then be checked to verify that the estimated L_c/r values are correct.

Finally, if we know the column load, we can use the following tables to select our member without additional calculation. The *Steel Construction Manual*[2] has greatly expanded tables for wide flange, hollow structural section, and pipe (Tables 6.3–6.5, respectively).

6.3 DEMAND VS CAPACITY

We have spent a lot of effort understanding buckling and column capacity, now we bring it all together. When the capacity is greater than the demand, our column has adequate strength. This relationship is given by Equation 6.12:

$$\phi P_n \geq P_u \tag{6.12}$$

Up to this point, we have looked at members with only one load type on them. However, often there is a combination of bending and axial load on a member, like the one in the tension example. Equations 6.13 and 6.14 govern this condition:

when $P_u/\phi P_n \geq 0.2$

$$\frac{P_u}{\phi P_n} + \frac{8}{9}\left(\frac{M_{ux}}{\phi M_{nx}} + \frac{M_{uy}}{\phi M_{ny}}\right) \leq 1.0 \tag{6.13}$$

when $P_u/\phi P_n < 0.2$

$$\frac{P_u}{2\phi P_n} + \left(\frac{M_{ux}}{\phi M_{nx}} + \frac{M_{uy}}{\phi M_{ny}}\right) \leq 1.0 \tag{6.14}$$

where

P_u = axial demand, either tension or compression, k (N)
ϕP_n = axial capacity, k (N)

M_{ux} = flexural demand about the x axis, k-in (N-mm)
ϕM_{nx} = flexural capacity about the x axis, k-in (N-mm)
M_{uy} = flexural demand about the y axis, k-in (N-mm)
ϕM_{ny} = flexural capacity about the y axis, k-in (N-mm).

This equation has its basis in the mechanics of materials, known as unity equations. If these are equal to or less than 1.0, our design is adequate.

Deformations are a combination of axial shortening from compression (or lengthening from tension) and bending deformations. An example of the net effect is shown in a foam member in Figure 6.11, where the axial and bending compression cause significant deformation, and the axial compression and bending tension negate each other.

This leads to the question of how a compression element can also develop moments even if it is not part of a moment frame. There are several ways in which moments can be developed within a typical gravity column. Moment can be developed due to eccentric loading (Pe), deflection of the end restraint for fixed based columns ($P\Delta$), and from deflections within the column ($P\delta$), illustrated in Figure 6.12. We will discuss this further in the next section.

6.4 DEFLECTION

For typical building columns, axial deflection is not a concern. However, in truss members and frames, it contributes to total deflection, as discussed in Section 3.4.

Lateral column deflections and eccentric loads can cause additional moments. Steel columns receive loading from beams attached to the

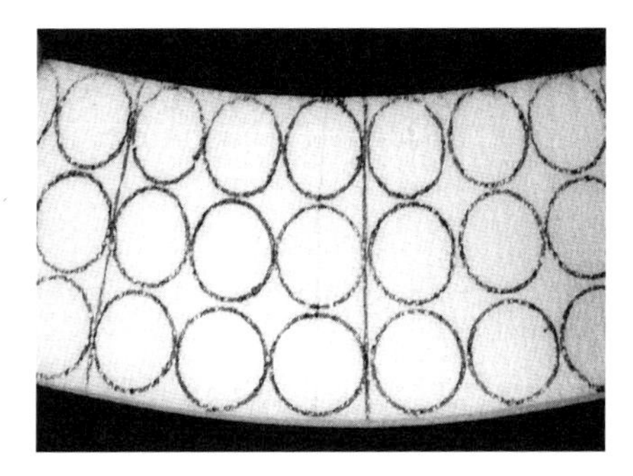

Figure 6.11 Combined compression and bending deformation in foam member

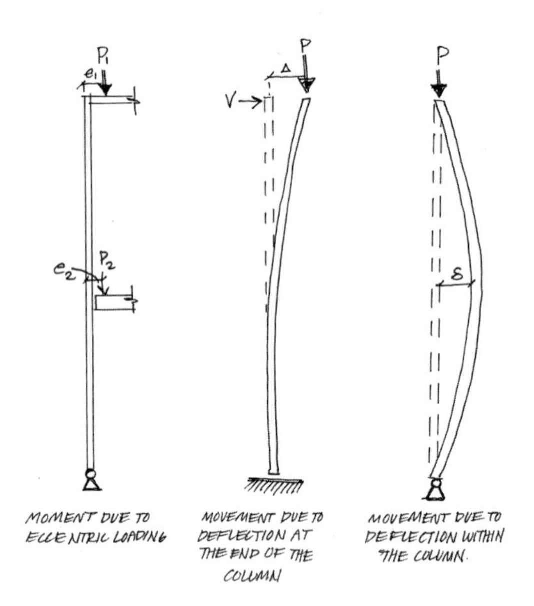

Figure 6.12 Additional moment sources in a column

column. If the column crosses several stories the beams at the lower floors connect to the side of the column (see Figure 6.12). This type of connection not only imparts axial load to the column but also creates a moment equal to the axial force times the distance to the row of bolts or welds. This moment is referred to as the eccentricity moment and is given by Equation 6.15:

$$M_u = P_u e \tag{6.15}$$

where

e = load eccentricity, ft (m).

The additional moment reduces the column capacity, seen when using the combined loading equations (6.13) and (6.14). For this reason, most connections at the top columns have the beams extending over the top. This delivers the load closer to the center of the column, minimizing the load eccentricity. This eccentricity not only creates its own moment, but helps to magnify the deflection, δ, within the column, which in turn increases the moment in the beam.

The other methods of developing moment are referred to as **P-Delta**. This is where the axial force within the column creates a moment due to the eccentricity created when the column deflects from the ideal straight line orientation. This can occur if the top of a cantilevered column deflects laterally, possibly due to a horizontal load. In this case delta is expressed as large delta, Δ as in Equation 6.16:

$$M_u = P_u \Delta \tag{6.16}$$

where

Δ = global load eccentricity, ft (m).

It can also occur if there is a deflection of the member between loads. In this case delta is expressed as small delta, δ, as in Equation 6.17:

$$M_u = P_u \delta \tag{6.17}$$

where

δ = local load eccentricity, ft (m).

P-delta moments are calculated based on the deflection by applying the initial forces. This is referred to as a first order analysis. However, when the moment is then added to the initial forces it results in an even larger deflection which in turn results in a larger P-delta moment. We therefore use an iterative analysis (second order) where the solution converges to stability or diverges to failure.

If a column is slender enough, it may appear acceptable from the first order analysis. However, the second order analysis, may show the increasing deflections and moments ultimately lead to failure.

6.5 DESIGN STEPS

To design a column, we must determine the following information:

Step 1: Determine the structural layout
Step 2: Determine the loads
Step 3: Determine material parameters
Step 4: Determine initial size
Step 5: Calculate capacity and compare to demand
- a. Global Buckling
- b. Local Buckling
- c. Torsional Buckling

Step 6: Summarize the Final Results

6.6 DESIGN EXAMPLE

Step 1: Determine the Structural Layout

We will design an interior steel column which carries gravity forces from the roof and floor framing, shown in Figure 6.13. The column runs for two stories, with the lower length of 22 ft (6.71 m). As the column is continuous without breaks over the entire height, we know the lower portion will control the design. Not only is the load greater, but the lower portion of the column has a larger unbraced length.

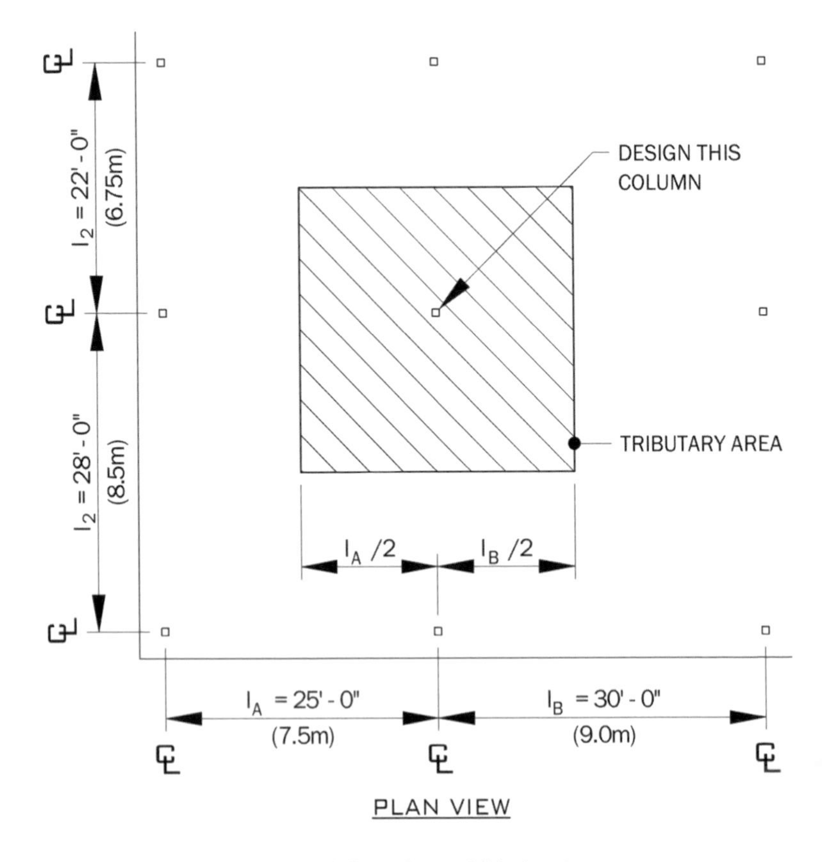

Figure 6.13 Column example (a) floor plan and (b) elevation

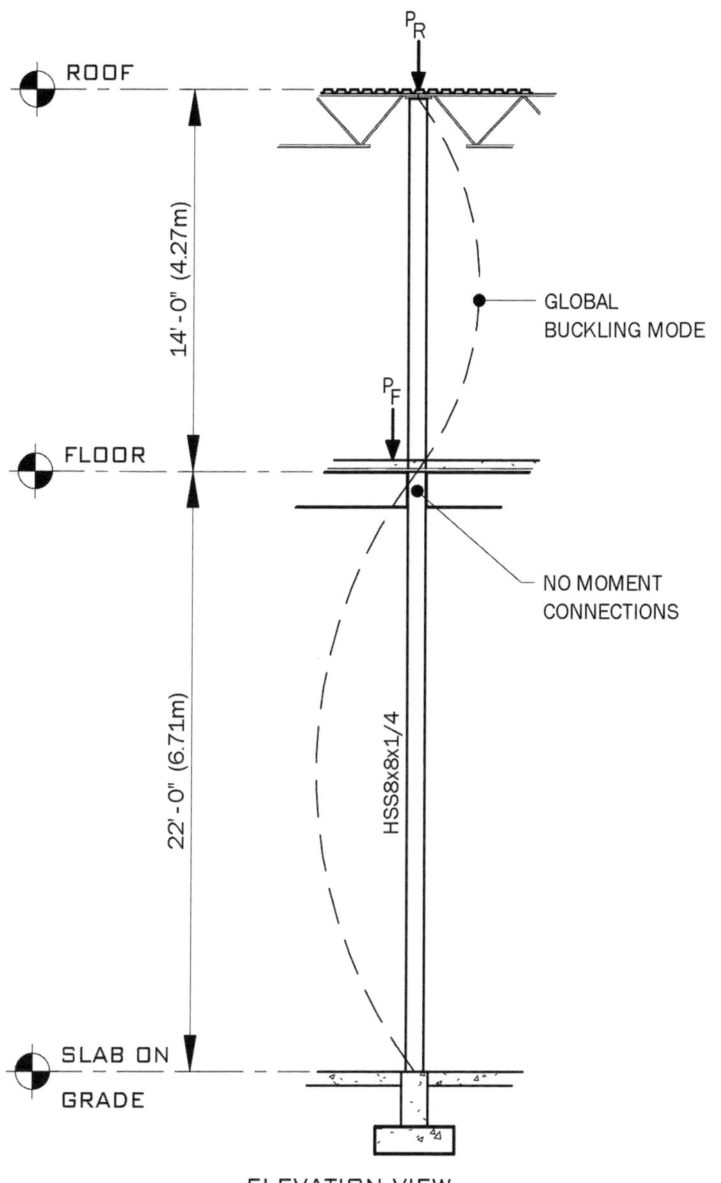
P_R
ROOF
14'-0" (4.27m)
GLOBAL
BUCKLING MODE
P_F
FLOOR
NO MOMENT
CONNECTIONS
22'-0" (6.71m)
HSS8x8x1/4
SLAB ON
GRADE
ELEVATION VIEW

Step 2: Determine the Loads

The roof unit loads are:

$q_D = 20$ lb/ft^2	$q_D = 0.960$ kN/m^2
$q_S = 45$ lb/ft^2	$q_S = 2.15$ kN/m^2

The floor unit loads are:

$q_D = 83$ lb/ft^2	$q_D = 4.00$ kN/m^2
$q_L = 100$ lb/ft^2	$q_L = 4.75$ kN/m^2

The code allows us to reduce the floor live load when the tributary area gets large enough. This is because the maximum live load will not occur everywhere at once. To apply the **live load reduction**, we need to determine the tributary area. Using Figure 6.13, we get

$$A_t = \left(\frac{L_A}{2} + \frac{L_B}{2}\right)\left(\frac{L_1}{2} + \frac{L_2}{2}\right)$$

$$= \left(\frac{25\text{ ft} + 30\text{ ft}}{2}\right)\left(\frac{28\text{ ft} + 22\text{ ft}}{2}\right) \qquad = \left(\frac{7.5\text{ m} + 9\text{ m}}{2}\right)\left(\frac{8.5\text{ m} + 6.75\text{ m}}{2}\right)$$

$$= 687.5\text{ ft}^2 \qquad = 62.91\text{ m}^2$$

Taking $K_L = 4$ for columns, the reduced live load follows:

$$L = L_o\left(0.25 + \frac{15}{\sqrt{K_{LL}A_T}}\right) \qquad L = L_o\left(0.25 + \frac{4.57}{\sqrt{K_{LL}A_T}}\right)$$

$$= 100\text{ lb/ft}^2\left(0.25 + \frac{15}{\sqrt{4(687.5\text{ ft}^2)}}\right) \qquad = 4.75\text{ kN/m}^2\left(0.25 + \frac{4.57}{\sqrt{(4)62.91\text{ m}^2}}\right)$$

$$= 53.6\text{ lb/ft}^2 \qquad = 2.56\text{ kN/m}^2$$

For a single floor, our live load must be at least 0.5 L_o, which it barely is.

Next, using the tributary area of the column, we can find the column point loads. At the roof,

$$P_{Dr} = 0.02\text{ k/ft}^2(687.5\text{ ft}^2) = 13.75\text{ k} \qquad P_{Dr} = 0.96\text{ kN/m}^2(62.91\text{ m}^2) = 60.4\text{ kN}$$

$$P_{Sr} = 0.045\text{ k/ft}^2(687.5\text{ ft}^2) = 30.94\text{ k} \qquad P_{Dr} = 2.15\text{ kN/m}^2(62.91\text{ m}^2) = 135.3\text{ kN}$$

At the floor,

$$P_{Df} = 0.083 \text{ k/ft}^2 \left(687.5 \text{ ft}^2\right) = 57.06 \text{ k}$$

$$P_{Dr} = 4.0 \text{ kN/m}^2 \left(62.91 \text{ m}^2\right) = 251.6 \text{ kN}$$

$$P_{Lf} = 0.054 \text{ k/ft}^2 \left(687.5 \text{ ft}^2\right) = 37.13 \text{ k}$$

$$P_{Lf} = 2.56 \text{ kN/m}^2 \left(62.91 \text{ m}^2\right) = 161.1 \text{ kN}$$

We next determine the maximum factored load. We will check the live and snow dominant combinations as follows:

$$\begin{aligned} P_u &= 1.2\text{D} + 1.6\text{L} + 0.5\text{S} \\ &= 1.2(13.75 \text{ k} + 57.06 \text{ k}) + 1.6(37.13 \text{ k}) + 0.5(30.94 \text{ k}) \\ &= 159.9 \text{ k} \end{aligned}$$

$$\begin{aligned} &= 1.2(60.4 \text{ kN} + 251.6 \text{ kN}) + 1.6(161.1 \text{ kN}) + 0.5(135.3 \text{ kN}) \\ &= 699.8 \text{ kN} \end{aligned}$$

$$\begin{aligned} P_u &= 1.2\text{D} + \text{L} + 1.6\text{S} \\ &= 1.2(13.75 \text{ k} + 57.06 \text{ k}) + 1.0(37.13 \text{ k}) + 1.6(30.94 \text{ k}) \\ &= 171.6 \text{ k} \end{aligned}$$

$$\begin{aligned} &= 1.2(60.4 \text{ kN} + 251.6 \text{ kN}) + 1.0(161.1 \text{ kN}) + 1.6(135.3 \text{ kN}) \\ &= 752 \text{ kN} \end{aligned}$$

It looks like snow combination controls, giving $P_u = 172$ k (752 kN).

Step 3: Determine Material Parameters

We will use a HSS section for the column, made of A500 Grade B steel, having the following strength:
$F_y = 46$ k/in^2 (315 N/mm^2).

Step 4: Determine Initial Size

We will do a quick calculation to get an idea of the initial size. Rearranging the equation:

$$\phi P_n = \phi F_{cr} A_g$$

to get the area, using a critical stress of 45% of the yield strength.

$$A_{est} = \frac{P_n}{F_{cr}} = \frac{P_n}{0.45F_y}$$

$$= \frac{172\text{ k}}{0.45(46\text{ k/in}^2)}$$

$$= 8.31\text{ in}^2$$

$$= \frac{752\text{ kN}}{0.45(315\text{ N/mm}^2)}\left(\frac{1000\text{ N}}{1\text{ kN}}\right)$$

$$= 5{,}305\text{ mm}^2$$

Based on this estimated area, we select an HSS6 × 6 × 3/8 (HSS152.4 × 152.4 × 9.5). From Table A1.4, the geometric properties we will need are

$A_g = 7.58\text{in}^2$ (4890 mm^2)
$r_y = r_x = 2.28$ in (58 mm).

Step 5: Calculate Capacity and Compare to Demand

Step 5A: Global Buckling

First, we calculate the slenderness ratio L_c/r, to determine whether the column is in the elastic or inelastic range. To determine L_c we need to find the effective length factor. For a column pinned at the top and bottom, $K = 1.0$, since the column is braced by the floor beams, but not moment connected. Thus,

$$L_c = KL$$

$$= 1.0(22\text{ ft})\left(\frac{12\text{ in}}{1\text{ ft}}\right)$$

$$= 264\text{ in}$$

$$= 1.0(6.71\text{ m})\left(\frac{1000\text{ mm}}{1\text{ m}}\right)$$

$$= 6{,}710\text{ mm}$$

and

$$\frac{L_c}{r} = \frac{264\text{ in}}{2.28\text{ in}} = 115.8$$

$$\frac{L_c}{r} = \frac{6{,}710\text{ mm}}{58\text{ mm}} = 115.7$$

We now compare this to

$$4.71\sqrt{\frac{E}{F_y}}$$

$$= 4.71\sqrt{\frac{29{,}000\text{ k/in}^2}{46\text{ k/in}^2}}$$

$$= 118.3$$

$$= 4.71\sqrt{\frac{200{,}000\text{ N/mm}^2}{315\text{ N/mm}^2}}$$

$$= 118.7$$

Since L_c/r is less than the limit, we know the capacity is controlled by inelastic buckling. First, we calculate the elastic buckling stress

$$F_e = \frac{\pi^2 E}{(L_c/r)^2}$$

$$= \frac{\pi^2 (29{,}000 \text{ k/in}^2)}{(115.8)^2} \qquad = \frac{\pi^2 (200{,}000 \text{ N/mm}^2)}{(115.7)^2}$$

$$= 21.34 \text{ k/in}^2 \qquad = 147.5 \text{ N/mm}^2$$

And now to find the critical buckling stress:

$$F_{cr} = \left(0.658^{F_y/F_e}\right) F_y$$

$$= \left(0.658^{46 \text{ k/in}^2/21.34 \text{ k/in}^2}\right) 46 \text{ k/in}^2 \qquad F_{cr} = \left(0.658^{315 \text{ N/mm}^2/147.46 \text{ N/mm}^2}\right) 315 \text{ N/mm}^2$$

$$= 18.66 \text{ k/in}^2 \qquad = 128.8 \text{ N/mm}^2$$

and finally the capacity:

$$\phi P_n = 0.9 F_{cr} A_g$$

$$= 0.9 (18.66 \text{ k/in}^2) 7.58 \text{ in}^2 \qquad = 0.9 (128.8 \text{ N/mm}^2) 4{,}890 \text{ mm}^2 \left(\frac{1 \text{ kN}}{1000 \text{ N}}\right)$$

$$= 127.3 \text{ k} \qquad = 567 \text{ kN}$$

The buckling capacity is less than the demand P_u, which is not OK. We need to go back to step 4 and choose a larger section.

Step 4: Determine Initial Size

We will now try an HSS8 × 8 × 1/4 (HSS203.2 × 203.2 × 6.4). Note that while the area is smaller, it has a larger radius of gyration. We'll see if this makes a difference. From Appendix 1, the section properties we will need are:

$A_g = 7.10 \text{in}^2$ (4580 mm²)
$r_y = r_x = 3.15$ in (80 mm).

Step 5: Calculate Capacity and Compare to Demand

Step 5A: Global Buckling

Again, we calculate the value for L_c/r and the elastic buckling stress:

$$\frac{L_c}{r} = \frac{264 \text{ in}}{3.15 \text{ in}} = 83.8 \qquad \frac{L_c}{r} = \frac{6{,}710 \text{ mm}}{80 \text{ mm}} = 83.9$$

$$F_e = \frac{\pi^2 (29{,}000 \text{ k/in}^2)}{(83.8)^2} = 40.7 \text{ k/in}^2$$

$$F_e = \frac{\pi^2 (200{,}000 \text{ N/mm}^2)}{(83.9)^2} = 280.4 \text{ N/mm}^2$$

And now the critical buckling stress follows:

$$F_{cr} = \left(0.658^{46 \text{ k/in}^2/40.7 \text{ k/in}^2}\right) 46 \text{ k/in}^2 = 28.7 \text{ k/in}^2$$

$$F_{cr} = \left(0.658^{315 \text{ N/mm}^2/280.4 \text{ N/mm}^2}\right) 315 \text{ N/mm}^2 = 196.9 \text{ N/mm}^2$$

The capacity is

$$\begin{aligned} \phi P_n &= 0.9 F_{cr} A_g \\ &= 0.9(28.7 \text{ k/in}^2) 7.1 \text{ in}^2 \\ &= 183.4 \text{ k} \end{aligned}$$

$$\begin{aligned} &= 0.9(196.9 \text{ N/mm}^2) 4{,}580 \text{ mm}^2 \left(\frac{1 \text{ kN}}{1000 \text{ N}}\right) \\ &= 811.6 \text{ kN} \end{aligned}$$

Because this is greater than the demand, we know our column works!

Step 5B: Local Buckling

Continuing on, we check local buckling. We begin by determining *b/t*

$$\begin{aligned} \frac{h}{t} &= \frac{h - 3t}{t} \\ &= \frac{8 \text{ in} - 3(0.233 \text{ in})}{0.233 \text{ in}} \\ &= 31.3 \end{aligned}$$

$$\begin{aligned} &= \frac{203 \text{ mm} - 3(5.92 \text{ mm})}{5.92 \text{ mm}} \\ &= 31.3 \end{aligned}$$

The limiting width-to-thickness limit for a hollow structural section is:

$$\lambda_p = 1.4 \sqrt{\frac{E}{F_y}}$$

For compact elements

$$= 1.4 \sqrt{\frac{29{,}000 \text{ k/in}^2}{46 \text{ k/in}^2}} = 35.2$$

$$= 1.4 \sqrt{\frac{200{,}000 \text{ N/mm}^2}{315 \text{ N/mm}^2}} = 35.3$$

Since the width-to-thickness ratio is less than the limiting value for slender elements, λ_p, we know local buckling doesn't reduce our strength. Very nice!

Step 6: Summarize the Final Results

In summary, our final column design is an HSS8 × 8 × 1/4 (HSS203.2 × 203.2 × 6.4), A500 Gr. B steel column.

6.7 WHERE WE GO FROM HERE

In this chapter, we have covered equations for column buckling of non-slender elements. When the engineer decides to design a column in the slender range, the capacity of the column can be determined using equations and direction that can be found in the AISC 360 specification.[3]

NOTES

1. *The Bible. Authorized King James Version*, Judges 16.21–30 (Oxford UP, 1998).
2. AISC, *Steel Construction Manual*, 14th Edition (Chicago: American Institute of Steel Construction, 2011).
3. AISC, *Specification for Structural Steel Buildings*, AISC 360 (Chicago: American Institute of Steel Construction, 2016).

Steel Lateral Design

Chapter 7

Paul W. McMullin

7.1 INTRODUCTION

Lateral loads on structures are commonly caused by wind, earthquakes, and soil **pressure**, and less commonly from human activity, waves, or blasts. These loads are difficult to quantify with any degree of precision. However, following reasonable member and system proportioning requirements, coupled with prudent detailing, we can build reliable steel structures that effectively resist lateral loads.

What makes a structure perform well in a windstorm is vastly different than an earthquake. A heavy, squat structure, such as the Parthenon in Greece can easily withstand wind—even without a roof. Its mass anchors it to the ground. On the other extreme, a tent structure could blow away in a moderate storm. Conversely, the mass of the Parthenon makes is extremely susceptible to earthquakes (remember earthquake force is a function of weight), while the tent in a seismic event will hardly notice what is going on.

Looking at this closer, wind forces are dependent on three main variables:

- Proximity to open spaces such as water or mud flats
- Site exposure
- Building shape and height.

In contrast, earthquake forces are dependent on very different variables:

- Nature of the seismic event
- Building weight
- Rigidity of the structural system.

Because we operate in a world with gravity forces, we inherently understand the gravity load paths of the simple building shown in Figure 7.1a.

Downward loads enter the roof and floors and make their way to the walls, columns and eventually **footings**. Lateral loads can take more time to grasp. But we can think of them as turning everything 90 degrees; the structure acting as a cantilevered beam off the ground, illustrated in Figure 7.1b.

The magnitude and distribution of lateral loads drives the layout of frames and **shear walls**. These resist lateral forces, acting like cantilevered beams poking out of the ground.

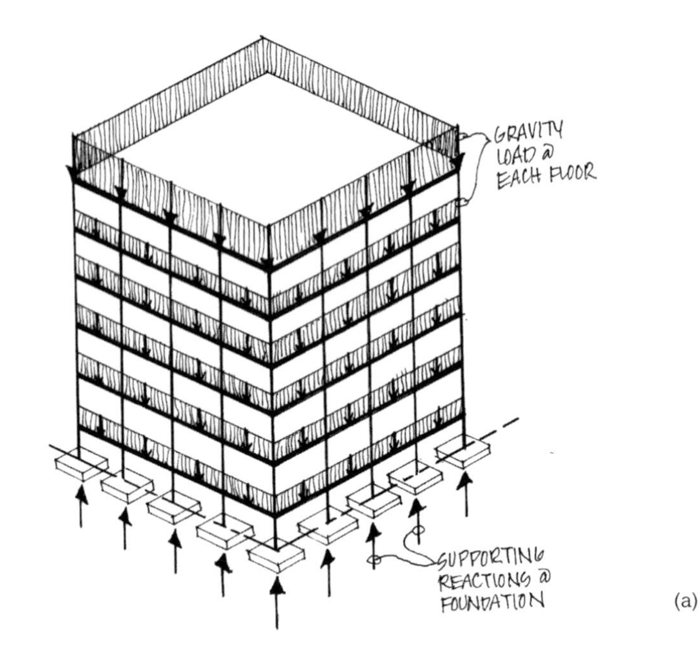

(a)

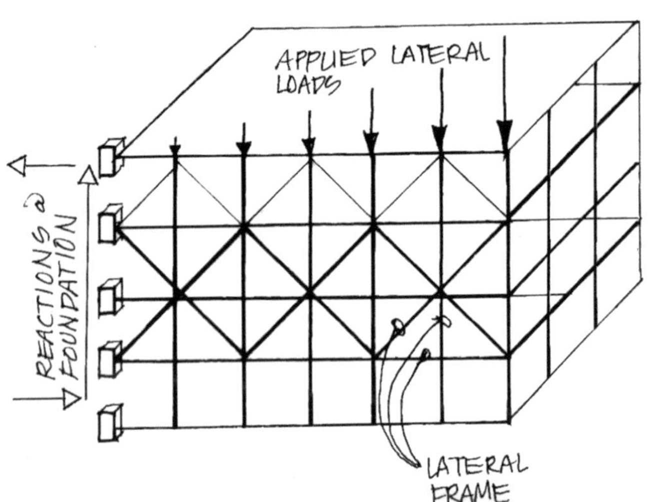

(b)

Figure 7.1 (a) Gravity load path, (b) lateral load path turned 90 degrees

We design lateral wind resisting members to not cause damage to the system. Conversely, because strong **seismic loads** occur much less frequently, we design their lateral systems to yield members. This absorbs significantly more energy, as illustrated in Figure 7.2, resulting in smaller member sizes. However, it leaves the structure damaged.

For design of **seismic load resisting systems**, we follow rigorous member proportioning and detailing requirements to ensure yielding occurs in the right places. This chapter focuses on design and detailing requirements from a conceptual point of view, and what lateral load resisting systems, elements, and connections should look like.

7.2 LATERAL LOAD PATHS

Following the path lateral loads travel through a structure is key to logical structural configuration and detailing. If the load path is not continuous from the roof to ground, failure can occur. Additionally, no amount of structural engineering can compensate for an unnecessarily complex load path.

When configuring the structure, visualize how lateral forces—and gravity forces—travel from element to element, and eventually to the ground. A well-planned load path will save weeks of design effort, substantially reduce construction cost, and minimize structural risk. Software can't do this, but careful thought will.

Looking at lateral load paths further, Figure 7.3 shows how they enter a structure and find their way to the ground. Starting at point 1, wind induces pressure, or seismic accelerations cause inertial forces, perpendicular to the face of the building. Spanning vertically (point 2),

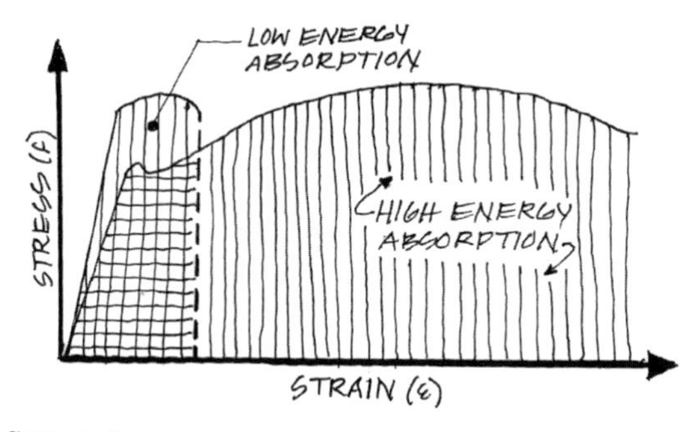

Figure 7.2 Comparative energy absorption for high and low deformation behavior

the wall delivers a line or point load to a connection at the roof or floor level. The roof or floor picks up additional inertial seismic load. The roof (number 3) must resist lateral forces through **diaphragm** action—essentially a deep beam. The ends of the diaphragm (point 4) then deliver load into connections between a shear wall or frame. This occurs at each level (point 5). The lateral force works its way to the footing (point 6), which transfers the force to the soil through friction and passive pressure. Because the lateral forces are applied at a distance above the ground, they impart an overturning moment to the system. This causes tension and compression in the ends of shear walls and outside frame columns (point 7). The weight of the structure (point 8) helps resist this overturning moment; keeping it from tipping over.

To review, lateral loads are applied perpendicular to walls or cladding. Bracing these are the roof and floor diaphragms, which transfer their

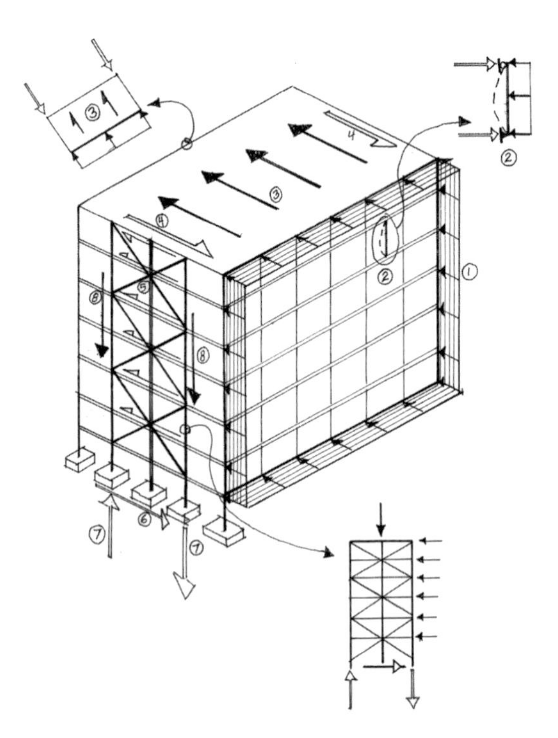

Figure 7.3 Detailed lateral load path in structure

loads to the walls parallel to the load. Walls are supported by the ground. The weight of the structure (and sometimes deep foundations) keeps the system from tipping over.

Connections are critical to complete load paths. We need to ensure the lateral loads flow from perpendicular wall and floor, into diaphragms, into walls parallel to the load, and down to the foundation. Each time the load enters a new element, there must be a connection.

7.3 DIAPHRAGMS

Lateral systems include horizontal and vertical elements. Horizontal systems consist of diaphragms and **drag struts** (**collectors**). Vertical elements consist of shear walls and frames. Horizontal systems transfer forces through connections to vertical elements, which carry the loads into the foundation.

Diaphragms may consist of concrete slabs, bare metal deck, and diagonal bracing. Diaphragms make possible large open spaces, without internal walls or braced frames—so long as there is adequate vertical support.

7.3.1 Forces

We can visualize diaphragms as deep beams that resist lateral loads, illustrated in Figure 7.4a. They experience maximum bending forces near their middle, and maximum shear at their supports (where they connect to walls or frames), as seen in Figure 7.4b.

We resolve the mid-span moments into a tension-compression **couple**, requiring boundary elements around their edges, such as beams. Often a few pieces of rebar in the slab can resist these forces, since the distance between these is large.

Shear forces are distributed throughout the length of the diaphragm in the direction of lateral force. Because many shear walls and frames do not go the length of the building, the transfer of shear forces between the diaphragm and vertical elements causes high stress concentrations at the ends of the wall or frame illustrated in Figure 7.5a. By adding drag struts (also called collectors), we gather the shear stresses into this stronger element, which can then deliver the force to the wall or frame. This reduces the stress concentration (Figure 7.5b) and ensures the diaphragm retains its **integrity**. Drag struts frequently consist of beams, joists, and slab reinforcing. Note that a structural element that acts as drag strut, will act as a diaphragm chord when the forces are turned and analyzed 90 degrees.

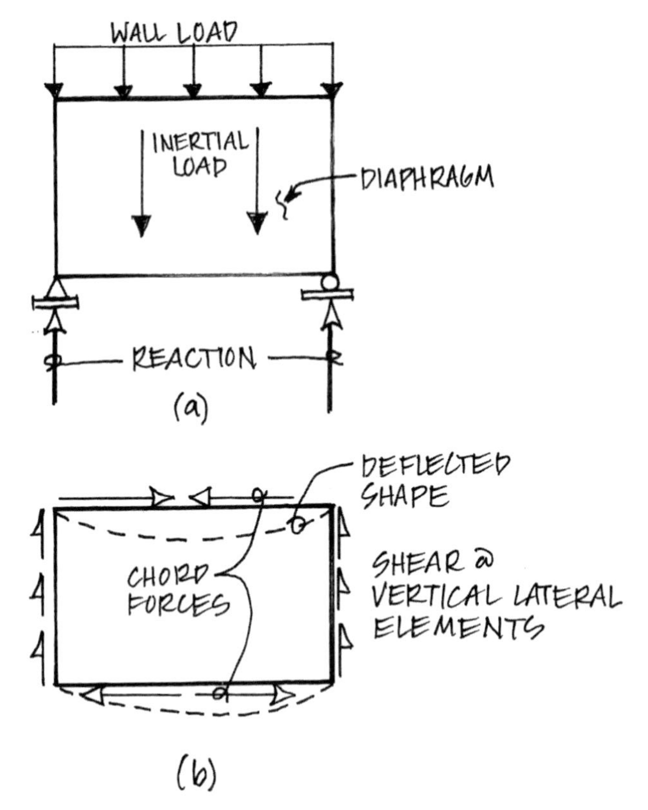

Figure 7.4 (a) Diaphragm forces and **reactions**, and (b) internal forces

7.3.2 Geometric Considerations

To ensure reasonable behavior of metal deck diaphragms, the ICC Evaluation Service reports[1] limit diaphragm **aspect ratios** (L/W) to those in Table 7.1. The limitations vary depending on whether the wall is flexible (curtain wall), or rigid (concrete or masonry). Generally, **flexibility factors** F are less than 70 for bare metal deck and less than 2 for concrete-topped diaphragms. We can use this table when laying out frame and shear walls, to ensure the diaphragms are well proportioned. Diaphragms that support concrete or masonry walls must meet additional span and deflection criteria.

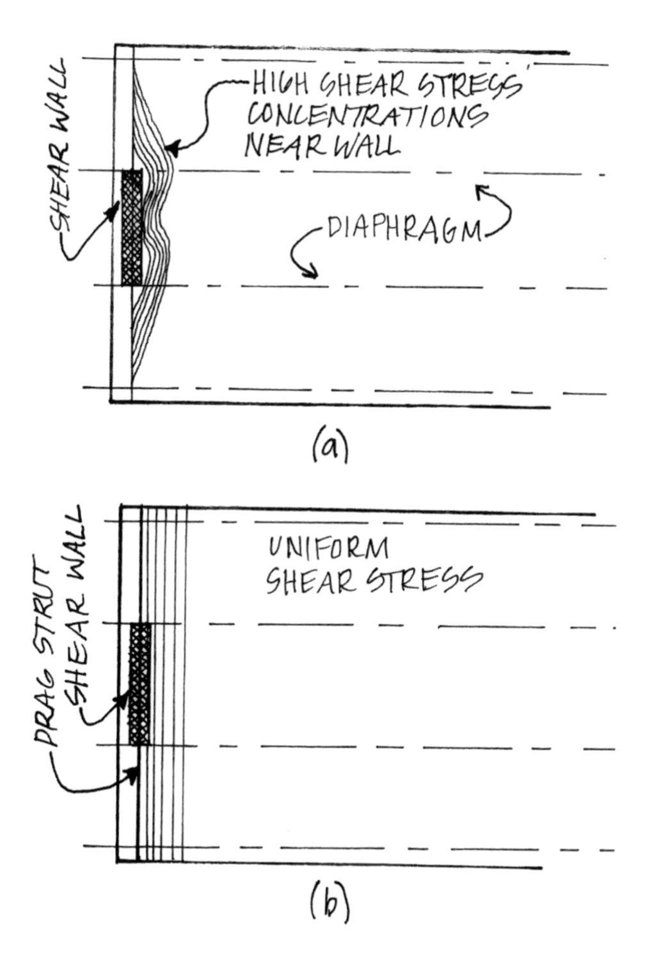

Figure 7.5 Diaphragm stress distribution (a) without and (b) with drag struts

7.3.3 Analysis

To design a diaphragm, we need to know the shear and moment distribution in it—though often just the maximum shear and moment. We take these and find the unit shear and tension-compression couple. The steps are as follows, illustrated in Figure 7.6.

- Draw the diaphragm and dimensions L and W
- Apply forces from the walls and floor as a line load w

Table 7.1 Metal deck diaphragm aspect ratio limits

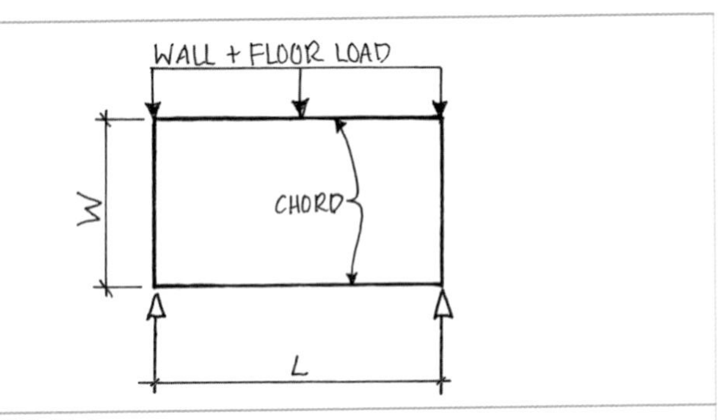

Span to Depth (L/W) Limits					
		Rotation not Considered		*Rotation Considered*	
Flexibility Factor F	*Max Span for Rigid Walls ft (m)*	*Rigid Walls*	*Flexible Walls*	*Rigid Walls*	*Flexible Walls*
>150	Not Permitted	Not Permitted	2:1	Not Permitted	1.5:1
70–150	200 (61)	2:1 or δ	3:1	Not Permitted	2:1
10–70	400 (122)	2.5:1 or δ	4:1	δ	2.5:1
1–10	No Limit	3:1 or δ	5:1	δ	3:1
<1	No Limit	δ	No Limit	δ	3.5:1

Source: ESR-1414

Notes

1) Flexibility factor F is a function of deck span, thickness, depth, and fastening. See ICC ESR 1414 for values
2) Rigid walls are those made of concrete or masonry
3) δ indicates additional deflection requirement

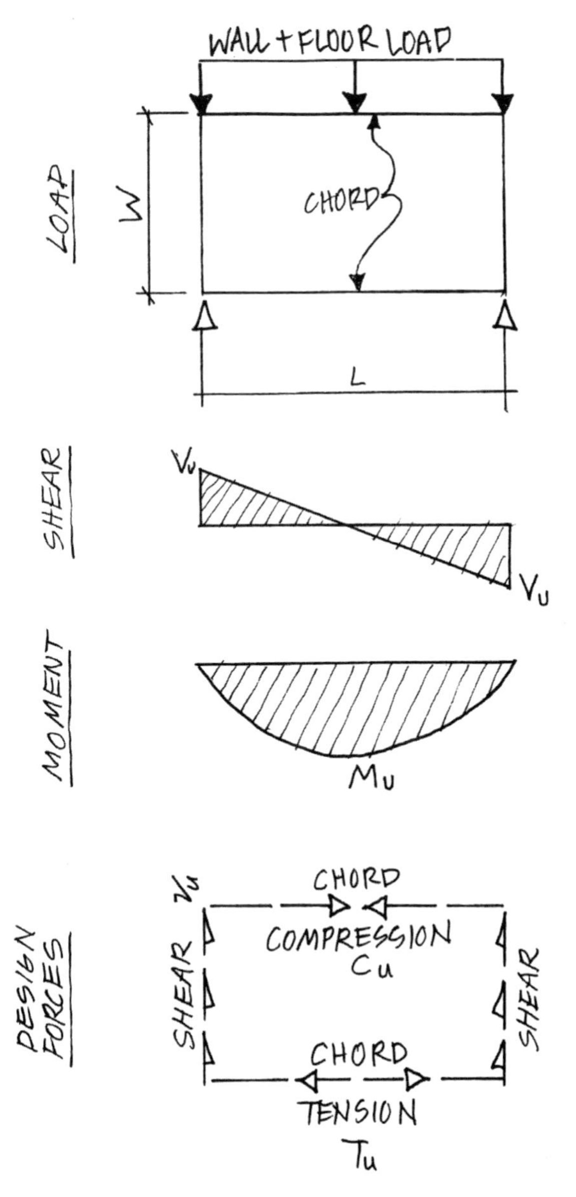

Figure 7.6 Diaphragm load, shear and moment diagram, and design forces

- Draw the shear and moment diagrams. For a simply supported diaphragm, the maximum shear and moment are shown in Equations 7.1 and 7.2:

$$V_u = \frac{w_u L^2}{2} \tag{7.1}$$

$$M_u = \frac{w_u L^2}{8} \tag{7.2}$$

where

w_u = uniform distributed load from walls and floors, lb/ft (kN/m)
L = span, ft (m).

- Calculate the unit shear v_u by dividing the shear force V_u by the depth W (Equation 7.3):

$$v_u = \frac{V_u}{W} \tag{7.3}$$

- Convert the moment M_u to a tension-compression couple as follows (Equation 7.4):

$$T_u = \frac{M_u}{W} \quad C_u = \frac{M_u}{W} \tag{7.4}$$

For the perpendicular direction, we follow the previous steps rotating the load and dimension labels 90 degrees.

7.3.3.1 Capacity

Knowing the diaphragm forces, we can size the chords. If we use beams we treat it as a combined **axial** and bending load, as discussed in Section 6.3. If we use reinforcing steel, we take the chord force and divide it by 0.9 F_y to get the required area, then select the necessary number of bars.

7.3.3.2 Detailing

Structural performance, particularly in earthquakes, depends on detailing. Figure 7.7 shows a typical building edge detail showing possible chords and shear transfer between the slab and beam. Figure 7.8 shows a chord detail across a column, where the beam-column connection doesn't have the capacity to carry the force.

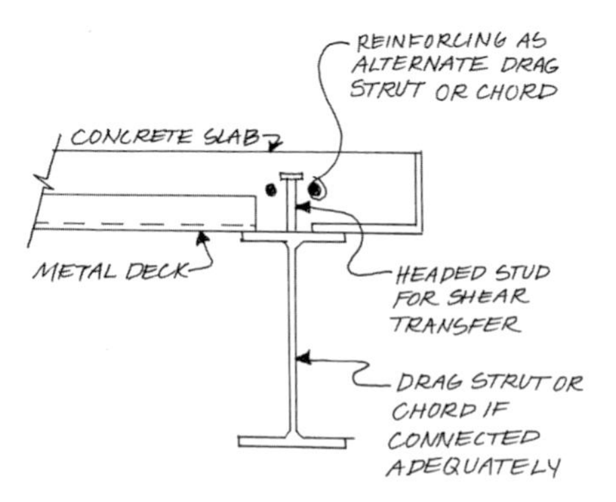

Figure 7.7 Detail of slab edge and chord

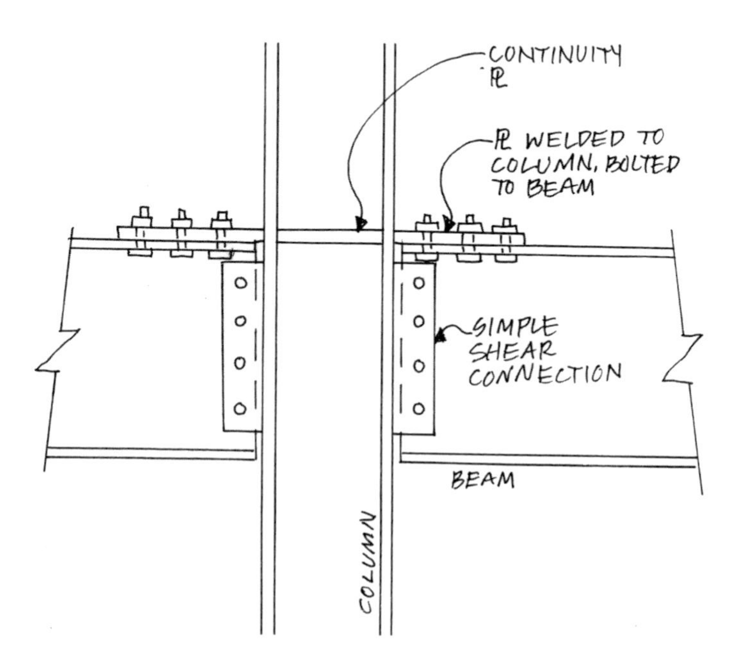

Figure 7.8 Chord splice across a column

7.4 LATERAL SYSTEM TYPES

There are as many variations in steel lateral systems as opinions on soccer clubs. The following sections discuss the most common: braced and moment frames. They can be configured in endless ways.

7.4.1 Braced Frames

Braced frames use axial strength and stiffness of braces, beams, and columns to resist lateral loads, illustrated in Figure 7.9. They act as cantilevered trusses. They are stiff and structurally efficient. However, they concentrate force to a few elements, which requires thoughtful consideration of connections and redundancy. The internal forces in a braced frame are illustrated in Figure 7.10.

The most common braced frame types are:

- Concentrically Braced Frames—where the loads are all transferred through the work points, and don't induce bending into the members. Seismic energy is absorbed by brace yielding.
- Eccentrically Braced Frames—where the braces do not meet at work points, and induce bending moments in the beams. In earthquakes, this causes the beams to yield, and dissipate energy.

Figure 7.9 Braced frame in architectural feature

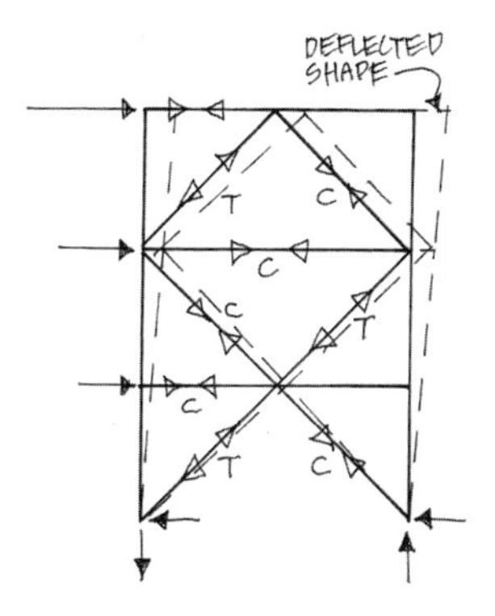

Figure 7.10 Braced frame internal forces and deflections

- Buckling Restrained Braced Frames—contain a steel core that yields equally in tension and compression. It is jacketed by concrete within a steel tube that keeps the core from yielding. They absorb seismic energy by core yielding.

There are many ways to configure braced frames. Figure 7.11 shows five configurations. The inverted V configuration is quite common, as it keeps the middle of the bay open for doors and windows. Flipping this, we get the V configuration. In seismic applications, these configurations create a large force imbalance, that the beam must carry. Two story braces eliminate this imbalance, and allow equally large openings. We configure **eccentric braced** frames with the link at the middle of the beam, or end,

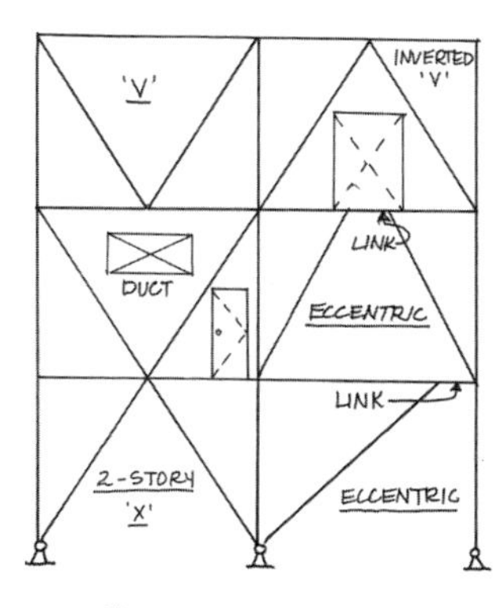

Figure 7.11 Braced frame configurations

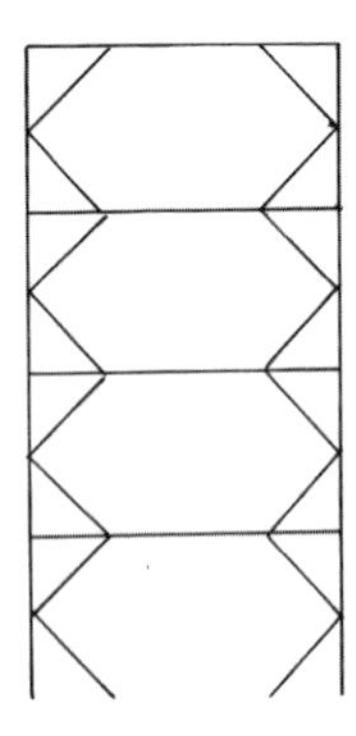

Figure 7.12 K-type bracing worth avoiding

shown in the bottom two bays of Figure 7.11. We prefer avoiding K type bracing, shown in Figure 7.12, which puts high, horizontal forces into the columns.

7.4.2 Moment Frames

Moment frames resist lateral loads through bending of beams and columns, with rigid connections between them. Internal forces and deflections of a moment frame are shown in Figure 7.13. Moment frames have high redundancy when compared to braced frames and shear walls, because of their many members and joints, shown in Figure 7.14.

Often moment frames are considered more expensive, when compared to braced frames. However, when we include the foundation costs in the analysis, they have similar costs. This is because moment frame foundations are lighter than braced frames, since they spread the load out over a greater area.

7.4.3 Dual Systems

We often use a combination of frame systems in buildings to resist lateral forces. For instance, core shear walls work well with typical office building layouts, and provide high strength and stiffness. When paired with perimeter moment or braced frames, the building has additional redundancy and torsional stiffness. When using dual systems, the code specifies **seismic response modification** factors R (see section 7.5.1) that consider the combined behavior of the two systems.

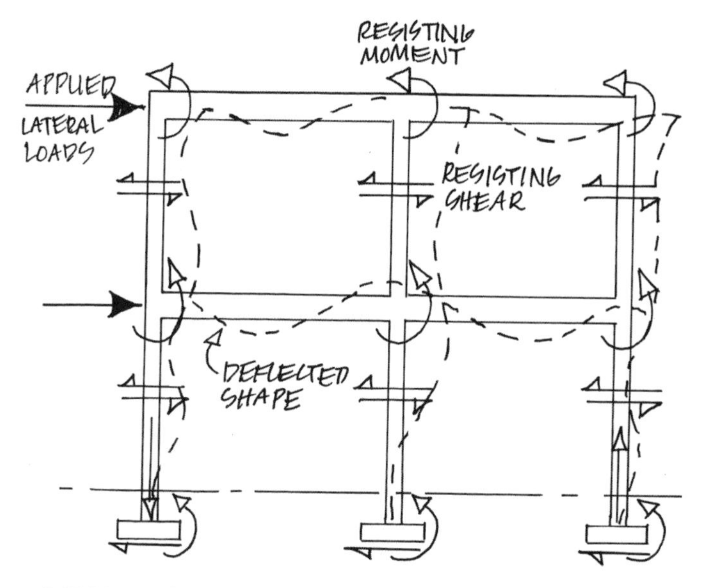

Figure 7.13 Moment frame internal forces and deformations

Figure 7.14 Moment frame in a commercial office building

7.5 SEISMIC DESIGN CONSIDERATIONS

Seismic design centers on yielding specific members in the structure to absorb energy. This reduces member sizes and creates more economical structures. In concentric braced frames, we yield the braces; in eccentric braced frames and moment frames, we yield the beams. We avoid yielding columns, as these contribute to the gravity load capacity and stability of much larger areas than their counterparts. Connections must be designed to yield these members, without failure. We therefore design them based on expected member strength, rather than forces from structural analysis.

After the 1994 Northridge earthquake, structural engineers learned that some of their design, detailing, and construction practices were not adequate to ensure yielding of specific elements. This event started a decade of research and modification to steel seismic requirements, now found in *AISC 341 Seismic Provisions for Structural Steel Buildings.*[2] These range from welding requirements to connection forces. We summarize key seismic provisions below.

7.5.1 Response Modification

Because we design seismic systems to yield, the building code permits us to reduce the design seismic force. We do this by dividing it by the Response Modification Factor R, which is a function of energy absorption. A higher R indicates a better performing seismic system. Table 7.2 provides these for various lateral force resisting systems.

Codes also limit the height of most lateral systems in high seismic regions. These limits are based on seismicity, which manifest themselves as **seismic design categories** B through F, and are listed in Table 7.2. Category D is the most common in regions of high seismicity. Categories E and F apply to very high seismicity, and buildings with higher societal importance.

7.5.2 Drift

For seismic forces, the code limits how much relative movement is permissible between floors—known as drift Δ. Think of it like a stack of dinner plates sliding off each other. Limiting drift helps reduce damage to cladding, partitions, mechanical ducts, and plumbing. Drift is determined from a structural analysis and compared to the limits shown in Table 7.3—which are a function of story height.

Drift in steel structures can have a substantial effect on how cladding joints are detailed. The joints need to accommodate the drift movement,

Table 7.2 Seismic lateral system *R* factors and maximum heights

	Response Coefficient	*Permitted Height (ft) Seismic Category*				
Seismic Force Resisting System	***R***	*B*	*C*	*D*	*E*	*F*
Timber						
Light Frame Walls Structural Panel Sheathed Walls	6 1/2	NL	NL	65	65	65
Concrete						
Special Moment Frames	8	NL	NL	NL	NL	NL
Special Reinforced Shear Walls	5	NL	NL	160	160	100
Steel						
Special Moment Frames	8	NL	NL	NL	NL	NL
Special Concentrially Braced Frames	6	NL	NL	160	160	100
Ordinary Concentrically Braced Frames	3 1/4	NL	NL	35	35	NP
Eccentrically Braced Frames	8	NL	NL	160	160	100
Buckling Restrained Braced Frames	8	NL	NL	160	160	100
Special Plate Shear Walls	7	NL	NL	160	160	100
Masonry						
Special Reinforced Shear Walls	5	NL	NL	160	160	100

Source: ASCE 7–10

NL= No Limit

NP= Not Permitted

without damaging the cladding. See Chapter 8 of *Special Structural Topics* in this series for additional guidance.

7.5.3 Configuration Requirements

Building configurations that have horizontal jogs, vertical steps, large diaphragm openings, or large stiffness changes perform less effectively than their counterparts. This is because force concentrates in sharp changes of geometry, and the load path through these is inefficient.

Table 7.3 Drift limits for multi-story structures

	Risk Category		
Structural System	*I or II*	*III*	*IV*
Structures 4 stories or less, non masonry, with interior walls & ceilings designed to accommodate drift	$0.025h_{sx}$	$0.020h_{sx}$	$0.015h_{sx}$
Masonry cantilever shear walls structures	$0.010h_{sx}$	$0.010h_{sx}$	$0.010h_{sx}$
Other masonry shear wall structures	$0.007h_{sx}$	$0.007h_{sx}$	$0.007h_{sx}$
All other structures	$0.020h_{sx}$	$0.015h_{sx}$	$0.010h_{sx}$

Source: ASCE 7–10

h_{sx}=Story height under level being considered, don't forget to convert to inches or mm

Examples of horizontal and vertical **irregularities** are shown in Figure 7.15 and Figure 7.16, respectively, along with potential options to avoid them.

Expanding further, horizontal structural irregularities include:

- Torsion occurs where there is a substantial difference in lateral system stiffness, such as a building with shear walls on three sides, with a moment frame on the fourth, illustrated in Figure 7.15.
- Reentrant Corners occur where there is an inside corner of the structure without frames or shear walls along them, shown also in Figure 7.15.
- Diaphragm Discontinuity happens where there are large openings in the diaphragms.

Table 7.3 ***continued***

Limit Criteria	*Allowable Drift Values for Various Criteria & Lengths*						
	Story Height (ft)						
	8	*10*	*12*	*14*	*15*	*16*	*20*
	Allowable Drift Δ_a (in)						
0.007 h_{sx}	0.67	0.84	1.01	1.18	1.26	1.34	1.68
0.010 h_{sx}	0.96	1.20	1.44	1.68	1.80	1.92	2.40
0.015 h_{sx}	1.44	1.80	2.16	2.52	2.70	2.88	3.60
0.020 h_{sx}	1.92	2.40	2.88	3.36	3.60	3.84	4.80
0.025 h_{sx}	2.40	3.00	3.60	4.20	4.50	4.80	6.00
Limit Criteria	*Story Height (m)*						
	2.5	*3.0*	*3.5*	*4.0*	*4.5*	*5.5*	*6.0*
	Allowable Drift Δ_a(mm)						
0.007 h_{sx}	17.5	21.0	24.5	28.0	31.5	38.5	42.0
0.010 h_{sx}	25.0	30.0	35.0	40.0	45.0	55.0	60.0
0.015 h_{sx}	37.5	45.0	52.5	60.0	67.5	82.5	90.0
0.020 h_{sx}	50.0	60.0	70.0	80.0	90.0	110	120
0.025 h_{sx}	62.5	75.0	87.5	100	113	138	150

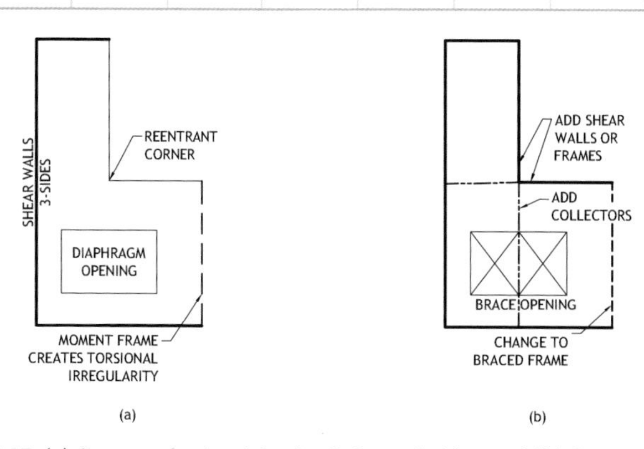

Figure 7.15 (a) Common horizontal seismic irregularities and (b) their mitigation

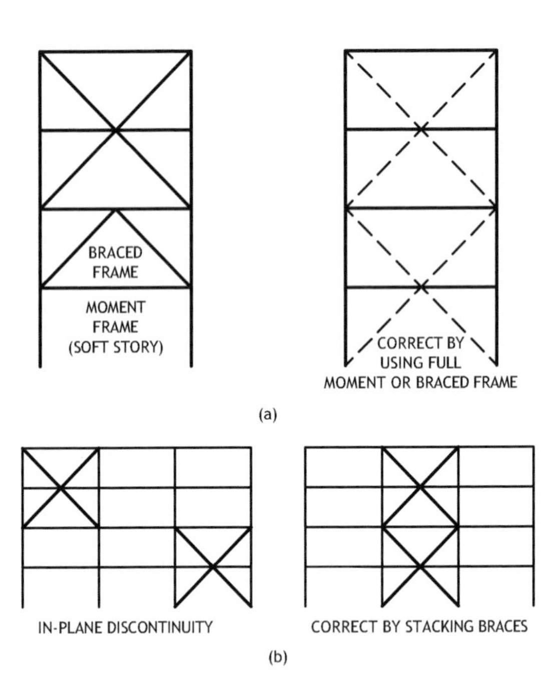

Figure 7.16 Vertical seismic irregularities showing (a) soft story and (b) in-plane discontinuities, and their mitigation

- Out-of-Plane Offsets occur where the lateral system changes plane and the forces must be transferred through the diaphragm to the vertical frames or shear walls.

Vertical structural irregularities include:

- Soft Stories occur where there is a drastic change in stiffness between levels. For example, a braced frame sitting on a moment frame, shown in Figure 7.16.
- Weak Stories exist where there is a large change in strength between levels.
- Mass irregularities occur where the adjacent story is 50% heavier than the adjacent stories.
- Geometric irregularities happen when the horizontal dimension of the lateral system changes more than 30% longer than an adjacent story.

- In-Plane Discontinuities occur where the lateral system changes locations horizontally, creating overturning forces in the members below, illustrated in Figure 7.16.

Each irregularity comes with specific, and sometimes exhaustive code requirements. Some of these are not permitted for seismic design categories D through F. Any lateral system with these irregularities will have financial and environmental costs, as these systems always require more material to carry the required loads. Additionally, no amount of analysis or detailing will make these structures perform as well as buildings without them.

7.5.4 Seismic Force Amplifications

There are a handful of cases where the code requires that the basic seismic forces be amplified. These include low redundancy conditions, structural irregularities, and protection of certain lateral frame elements.

Redundancy is the ability of a structure to sustain damage without becoming unstable. If failure of a key element of the lateral force resisting system results in a reduction of story shear strength of greater than 33%, the seismic force must be increased by 30%. More, smaller frames or walls will result in a less expensive, better performing seismic system.

When we have structural irregularities, as discussed in the previous section, we must increase the forces in elements that will be affected by these. The amount depends upon the irregularity and member, but ranges from 25% to 300%.

7.5.5 Material Requirements

As a class of steel becomes stronger it loses deformation capacity and fracture toughness. AISC therefore limits the yield stress of structural steels used in seismic applications to 50 k/in^2 (345 MN/m^2) for intermediate and **special** systems, and 55 k/in^2 (379 MN/m^2) for ordinary systems.

Heavy sections and welds have **Charpy** toughness requirements to ensure they can absorb sufficient seismic energy. Heavy sections are defined as rolled shapes with flanges thicker than 1 1/2 in (38 mm) and plate 2 in (51 mm) and thicker.

Table 7.4 Representative wide flange shapes that qualify as highly ductile for use in seismic systems

Section (Imperial)	*SMF*	*SCBF*		*EBF*		$P_{u\,ma}$		*Section (Metric)*
	Beam	*Brace*	*Beam*	*Brace*	*Beam*	*Highly Ductile*		
	Col		*Col*		*Col*	*(k)*	*(kN)*	
W44 × 335	■		■		■	3,900	(17,348)	W1100 × 499
W40 × 593	■	■	■	■	■	NL		W1000 × 883
W40 × 397	■	■	■	■	■	NL		W1000 × 584
W40 × 149	■		■		■	169	(752)	W1000 × 222
W36 × 652	■	■	■	■	■	NL		W920 × 970
W36 × 210	■		■		■	2,290	(10,186)	W920 × 313
W36 × 135								W920 × 201
W33 × 387	■	■	■	■	■	NL		W840 × 576
W33 × 201	■		■		■	1,810	(8,051)	W840 × 299
W33 × 118								W840 × 176
W30 × 391	■	■	■	■	■	NL		W760 × 582
W30 × 211	■	■	■	■	■	NL		W760 × 314
W30 × 90								W760 × 134
W27 × 539	■	■	■	■	■	NL		W690 × 802
W27 × 217	■	■	■	■	■	NL		W690 × 323
W27 × 84								W690 × 125
W24 × 370	■	■	■	■	■	NL		W610 × 551
W24 × 103	■		■		■	1,110	(4,938)	W610 × 153
W24 × 55	■		■		■	58.5	(260)	W610 × 82
W21 × 201	■	■	■	■	■	NL		W530 × 300
W21 × 122	■	■	■	■	■	NL		W530 × 182
W21 × 44	■		■		■	57.6	(256)	W530 × 66
W18 × 311	■	■	■	■	■	NL		W460 × 464

Table 7.4 *continued*

Section (Imperial)	*SMF*	*SCBF*		*EBF*		$P_{u\,ma}$		*Section (Metric)*
	Beam	*Brace*	*Beam*	*Brace*	*Beam*	*Highly Ductile*		
	Col		*Col*		*Col*	*(k)*	*(kN)*	
W18 × 158	■	■	■	■	■	NL		W460 × 235
W18 × 50	■		■		■	326	(1,450)	W460 × 74
W16 × 100	■	■	■	■	■	NL		W410 × 149
W16 × 57	■	■	■	■	■	NL		W410 × 85
W16 × 26								W410 × 38.8
W14 × 730	■	■	■	■	■	NL		W360 × 1086
W14 × 257	■	■	■	■	■	NL		W360 × 382
W14 × 43								W360 × 64
W12 × 336	■	■	■	■	■	NL		W310 × 500
W12 × 50	■	■	■	■	■	NL		W310 × 74
W12 × 26								W310 × 38.7
W10 × 112	■	■	■	■	■	NL		W250 × 167
W10 × 49								W250 × 73
W10 × 33								W250 × 49.1
W8 × 67	■	■	■	■	■	NL		W200 × 100
W8 × 24								W200 × 35.9

Source: AISC Seismic Design Manual, 2nd Edition

■ indicates highly ductile section

Blank indicates section insufficiently compact

NL no limit

7.5.6 Member Compactness

AISC 341 requires that members meet compactness criteria to ensure sufficient deformation in the members—and therefore sufficient energy dissipation. They define two ductility levels, moderate and high, which correlate to limiting width to thickness ratio (b/t) and web length to thickness ratio (h/t_w). These are like those discussed in Section 2.4.2, but more restrictive. Ordinary and Intermediate seismic detailing requires the members to meet moderate ductility requirements, while those in the Special category must meet the highly ductile requirements. Boiling these requirements down, we see a limited number of available shapes for seismic systems. Table 7.4 and Table 7.5 lists shapes from Appendix 1, that qualify as highly ductile seismic sections for select lateral systems.

7.5.7 Protected Zone

To ensure large deformation capacity, and therefore energy absorption, certain portions of seismic connections and members are considered **protected zones**, shown in Figure 7.17. No connections or notches can be made in these areas, including connections for non-structural items.

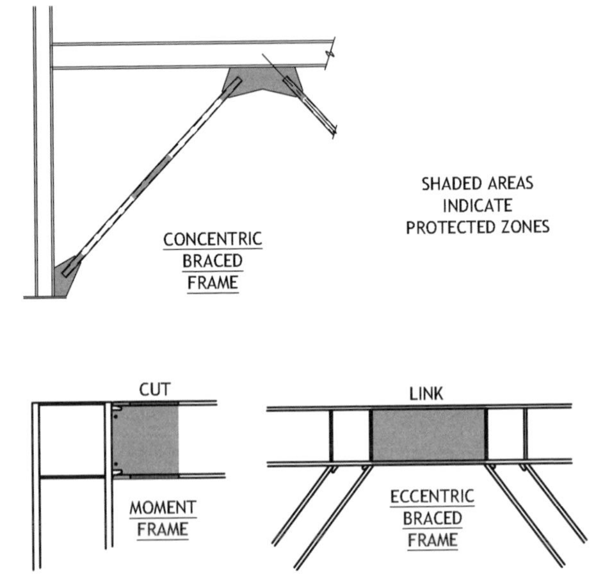

Figure 7.17 Seismic protected zones for (a) concentric braced frames, (b) eccentrically braced frames, and (c) moment frames

Table 7.5 Representative HSS sections that qualify as highly ductile for use in seismic systems

Section (Imperial)	*SCBF Brace*	*EBF Brace*	*Section (Metric)*
HSS16 × 16 × 3/8			HSS406.4 × 406.4 × 9.5
HSS14 × 14 × 3/8			HSS355.6 × 355.6 × 9.5
HSS12 × 12 × 3/8			HSS304.8 × 304.8 × 9.5
HSS10 × 10 × 5/8		■	HSS254 × 254 × 15.9
HSS10 × 10 × 3/8			HSS254 × 254 × 9.5
HSS8 × 8 × 5/8	■	■	HSS203.2 × 203.2 × 15.9
HSS8 × 8 × 3/8			HSS203.2 × 203.2 × 9.5
HSS7 × 7 × 5/8	■	■	HSS177.8 × 177.8 × 15.9
HSS7 × 7 × 3/8			HSS177.8 × 177.8 × 9.5
HSS7 × 7 × 1/4			HSS177.8 × 177.8 × 6.4
HSS6 × 6 × 5/8	■	■	HSS152.4 × 152.4 × 15.9
HSS6 × 6 × 3/8			HSS152.4 × 152.4 × 9.5
HSS6 × 6 × 1/4			HSS152.4 × 152.4 × 6.4
HSS5 × 5 × 3/8	■	■	HSS127 × 127 × 9.5
HSS5 × 5 × 1/4			HSS127 × 127 × 6.4
HSS4 × 4 × 3/8	■	■	HSS101.6 × 101.6 × 9.5
HSS4 × 4 × 1/4			HSS101.6 × 101.6 × 6.4
HSS3 × 3 × 1/4	■	■	HSS76.2 × 76.2 × 6.4
HSS3 × 3 × 1/8			HSS76.2 × 76.2 × 3.2

Source: AISC Seismic Design Manual, 2nd Edition

■ indicates highly ductile section

Blank indicates section insufficiently compact

7.5.8 Connections

Good detailing is at the heart of safely performing buildings. For wind forces, we design the connections for the actual wind force. However, for seismic connections, we must ensure a part of the structure yields. This leads to larger connections and more demanding material requirements.

We typically design members using the lowest strength they will have coming from the mill—F_y and F_u. In practice, the material may be stronger. This is usually helpful until we design seismic connections, which must force the member to yield. If the actual strength is higher than we expect, the required force increases, and our connection may be undersized. To account for this, we factor the yield and ultimate strengths up by the **expected strength** factors when calculating connection demand. Yield strength is multiplied by R_y and ultimate strength by R_t. These values are provided in Table 7.6.

For concentric braced frames, yielding occurs in the braces and we design the connections to develop the tensile capacity of the brace. This leads to rather large connections, such as those in Figure 7.18.

Table 7.6 Expected strength adjustment factors

ASTM Specification	*Yield* R_y	*Ultimate* R_t
Structural Shapes and Bars		
A36	1.5	1.2
A53	1.6	1.2
A500	1.4	1.3
A572 Gr 50	1.1	1.1
A913	1.1	1.1
A992	1.1	1.1
A588	1.1	1.1
Plate		
A36	1.3	1.2
A572 Gr 50	1.1	1.2

Source: AISC 341–10

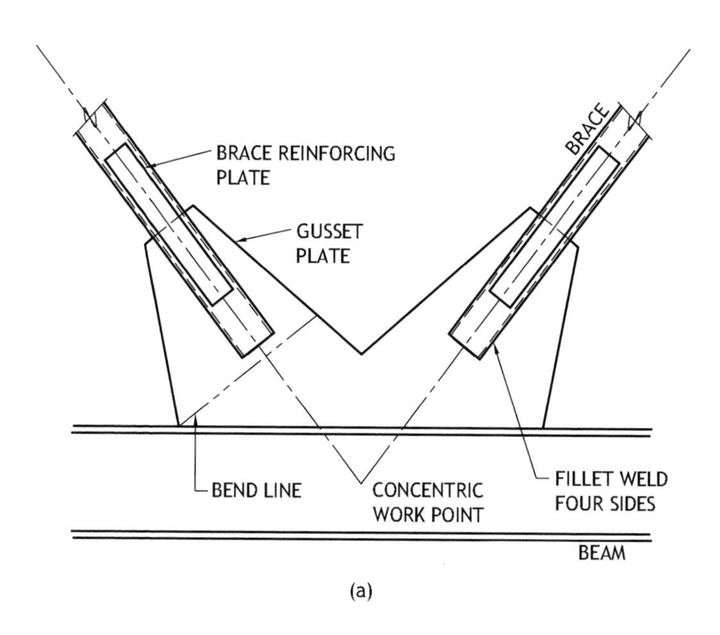

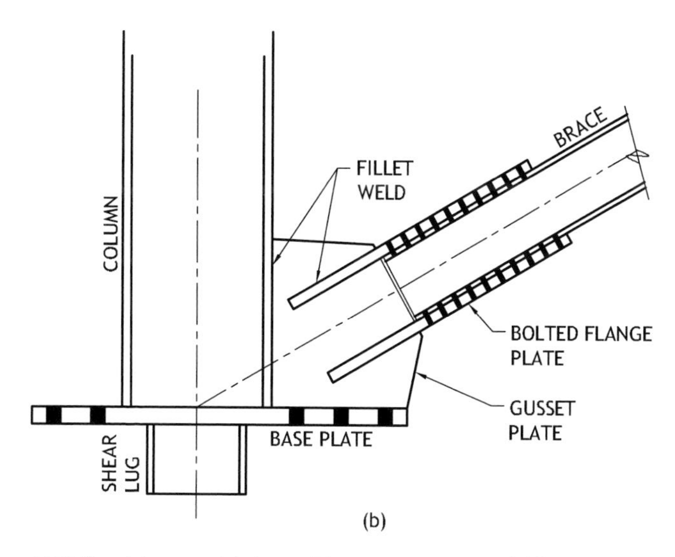

Figure 7.18 Special concentric braced frame connection at (a) beam mid-span, and (b) base plate

In eccentric braced frames the yielding occurs in the link beam, between braces. Such a connection is shown in Figure 7.19. For moment frames, we proportion the beam to yield. Common seismic moment connections are shown in Figure 7.20. In the bolted connection, the flange plate and bolts are sufficient to yield the beam in bending. In the welded connection, the beam section is reduced to yield, but keep the weld stresses elastic. In both, the shear connection in the web must be able to develop the expected strength of the beam, at the opposite end.

Bolts in seismic connections must be slip critical. This is to reduce drift and increase energy dissipation.

For welds, an entire book can be written on their requirements, and indeed has. *AWS D1.8 Structural Welding Code–Seismic Supplement*[3] provides the US code requirements for welds in seismic systems—also known as demand critical welds.

A key to successful welds is the access hole where the flange meets the web of a wide flange shape, shown in Figure 7.21. Proper configuration of this hole reduces constraint in the joint and allows the welder to properly make the weld.

7.6 WHERE WE GO FROM HERE

This chapter has introduced the general concepts of lateral design. From here, we estimate lateral forces on a structure, and through structural analysis, determine their distribution into diaphragms, frames, and shear walls. This yields internal forces, from which we proportion

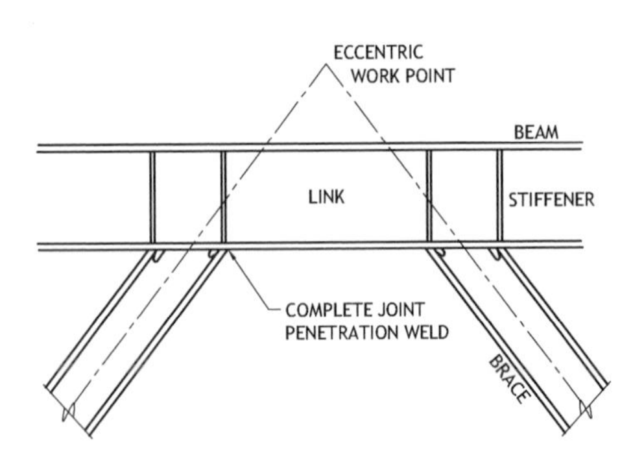

Figure 7.19 Eccentrically braced frame connection detail

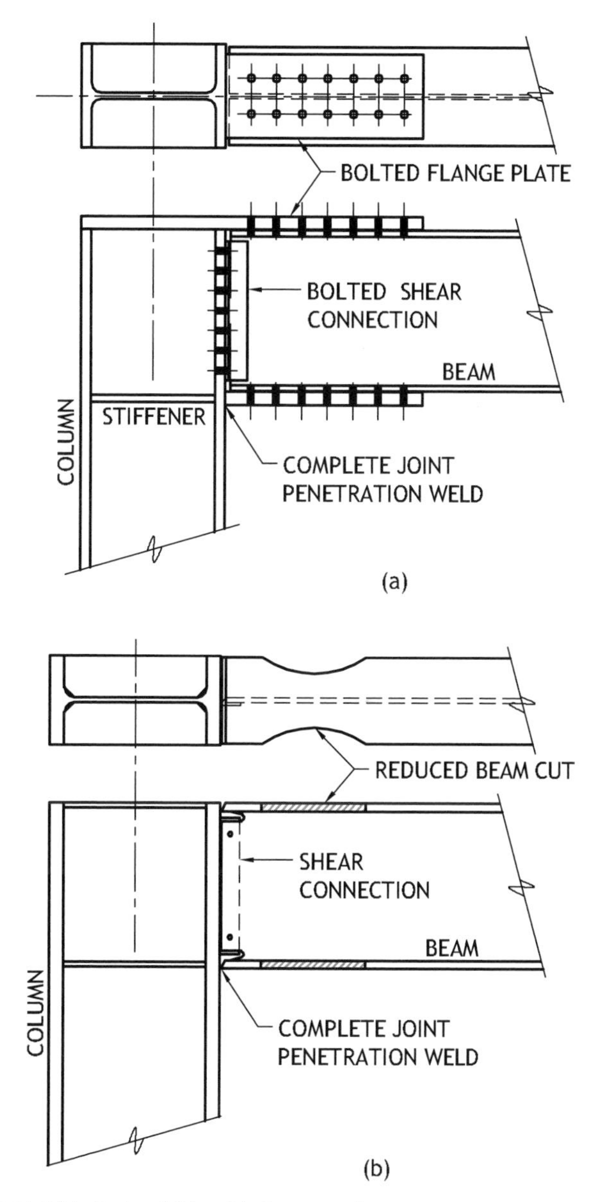

Figure 7.20 (a) bolted and (b) welded moment frame connection

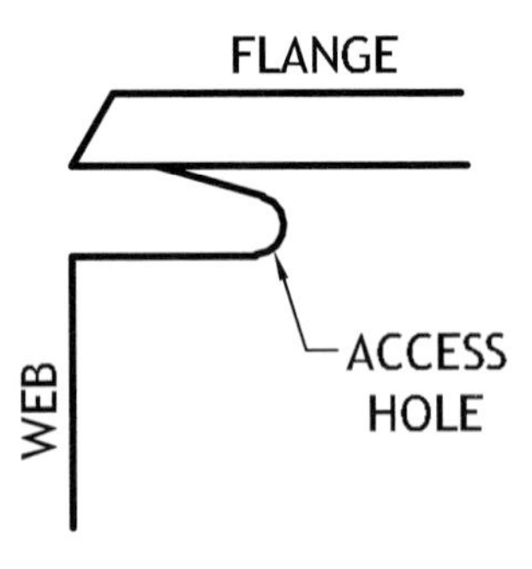

Figure 7.21 Seismic weld access hole geometry

member sizes. We then detail the structure, paying attention to the seismic requirements discussed above.

Seismic lateral design has become increasingly sophisticated in the past two decades. Prescriptive code requirements are giving way to performance-based design (PBD). This allows the owner and designer to pair the earthquake magnitude and structural performance that is consistent with the function of the building. Additionally, engineers are using performance-based design for more traditional, code-based buildings to reduce material consumption, as discussed in the *Special Structural Topics* volume of this series.

NOTES

1. ICC-ES, *Vulcraft Steel Deck Panels*, ESR-1227 (Brea, CA; ICC Evaluation Service, 2016)
2. AISC, *Seismic Provisions for Structural Steel Buildings*, AISC 314 (Chicago: Research Council on Structural Connections, 2016).
3. AWS, *Structural Welding Code-Seismic Supplement*, AWS D1.8 (Miami: American Welding Society, 2016).

Steel Connections

Chapter 8

Paul W. McMullin

Figure 8.1 Braced frame connection showing bolts, welds, and connected elements

Bolts, welds, and plates, like those shown in Figure 8.1, hold it all together. They transfer load between beams, columns, braces, chords, webs, and foundations to complete a structure. They allow individually fabricated pieces of steel to soar to the sky or span great lengths. In this chapter, we will learn the fundamentals of bolts, welds, and connected elements; looking at their properties, strength, detailing, and installation.

Regardless of the connector type, we need to ensure a continuous load path between the members being connected. Recalling Figure 3.5, we can look at a connection like a chain. Each time load moves from one element to another, there must be sufficient capacity. In the connection in Figure 8.1, the force moves from the beam through the weld into the **gusset**, then into the bolt, into the brace connection with the help of another weld, and finally the brace body. The analysis can be tedious, but necessary for a functional connection.

8.1 CONNECTOR TYPES

8.1.1 Bolts

In the first steel structures, connections were made of **rivets**, like those in Figure 8.2. Workers installed them red hot, and pounded them into shape, filling the hole. As the rivets cooled and contracted, they clamped the surfaces tightly together. Load transfer occurred by bearing and friction. As rivets were time consuming and dangerous to install, in the 1950s **high-strength bolts** found their way into use, and have completely replaced rivets.

Figure 8.2 Riveted connection in an industrial structure

There are two major classes of high-strength bolted connections—bearing and slip critical. Bearing bolts transfer force by contact between the plate and bolt, illustrated in Figure 8.3a. In slip critical joints, load is transferred through friction between the plates, with the bolt applying the clamping (normal) force (Figure 8.3b). Bearing bolts are common in smaller structures. Slip critical bolts are required in taller structures (over 125 ft, 38 m), bolted seismic systems, and those subject to **fatigue** and vibration.

Today, bolts are divided into three main divisions: Groups A and B, and A307. Groups A and B are high-strength bolts for structural connections. A307 bolts are what we commonly get at the hardware store, and are found in light connections like girts and handrails. Group A bolts include A325 and F1852. They have similar strength to SME J429[1] Grade 5 bolts. Group B bolts include A490 and F2280, and have similar strength as Grade 8 bolts. Availability of different bolt materials and their use are shown in Table 2.6. Bolts are identified by markings on their heads, like

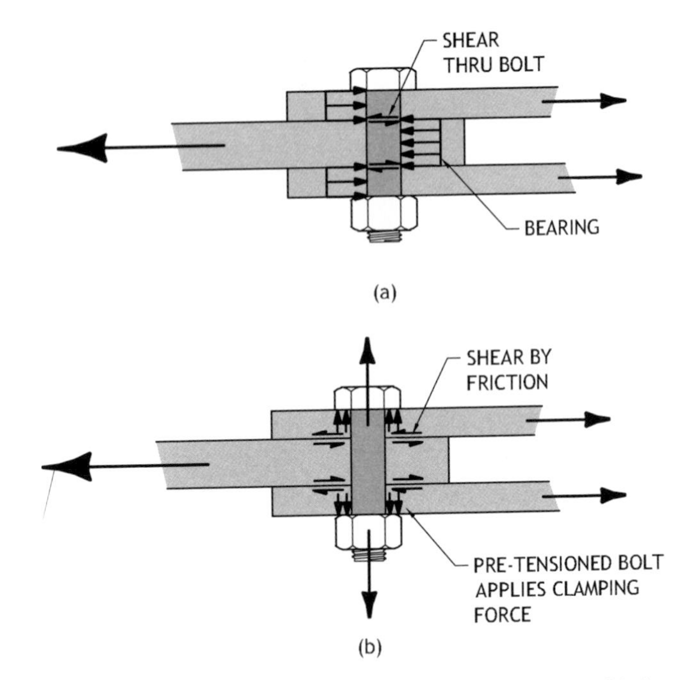

Figure 8.3 Force transfer in (a) bearing-type bolts, and (b) slip critical bolts

those in Figure 8.4. Their use is governed by the *Specification for Structural Joints using High-Strength Bolts.*[2]

8.1.2 Welds

The earliest record of welding dates to 5500 BC, where the Egyptians made copper pipe from sheets by overlapping the edges and hammering them together. This is known as **forge welding**, and was utilized to create the intricate patterns in Damascus steel. **Brazing** became widespread by 3000 BC, and utilizes a **filler metal** with a lower melting point than the material it is joining. It is still used widely today to join pipes and manifolds, where high structural strength is not required.

Arc welding development began independently in the early 1800s by Sir Humphry Davy and Vasily Petrov. Loosely paralleling advances in electrical generation, arc welding development really took off in the last two decades of the 19th century and early 20th century. By World War I (1914), welding was used extensively in ship repair. In 1924, the first all-

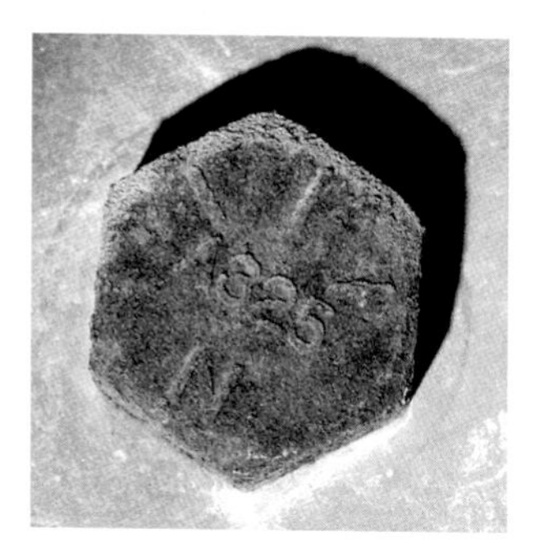

Figure 8.4 Bolt heads showing identifying marks

welded steel building was built in the US. Today welding is widespread in commercial, industrial, and bridge structures throughout the world.

Along with the development of welding came the need for standards to ensure the constructed product was properly built. The American Welding Society produced the first structural welding code in 1928. Since that time, they have developed the codes in the USA governing welding from reinforcing steel to aerospace components. The key welding codes for building and bridge structures are:

- D1.1 Structural Welding Code—Steel[3]
- D1.4 Structural welding (reinforcing steel)[4]
- D1.5 Bridge Welding Code[5]
- D1.8 Structural Welding Code—Seismic.[6]

Many arc welding processes exist today, yet they share common elements illustrated in Figure 8.5.

These are:

- Base Metals—to be joined
- Filler Metals—to fuse with the base metals into a solid weldment
- Electric Circuit—to supply the electrical energy needed create an electric arc to melt the metals
- Electrodes—to focus the electrical energy into the base metals
- Shielding Media—flux and/or shielding gases to protect the molten metal from atmospheric contamination
- Welding Power Sources—to create or rectify current and precisely control the voltage and the amperage flowing through the welding circuit.

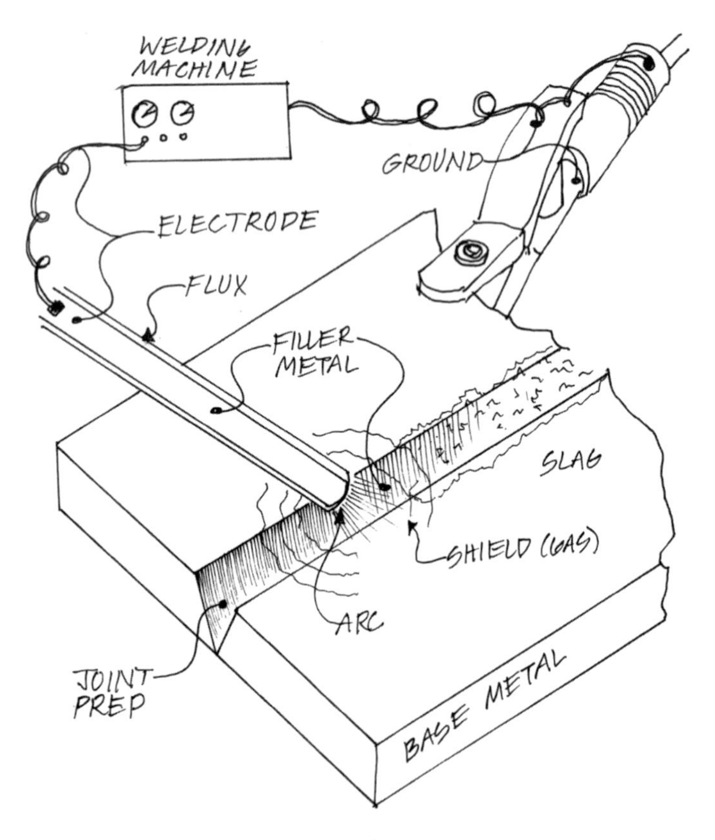

Figure 8.5 Required elements for arc welding

Common welding methods today include:

- Shielded Metal Arc Welding (SMAW)—commonly called "stick" welding, utilizes a flux-coated metal wire that acts both as an electrode and as the filler metal (Figure 8.5). The flux melts with the wire in the heat of the welding arc. The shielding gas and slag that are created protect the weld. Additional alloying elements are added to the flux to provide specific operating and metallurgical characteristics. It is the most versatile and accessible arc welding process.
- Submerged Arc Welding (SAW)—employs a continuous wire electrode that melts in the arc beneath a bed of granular flux. SAW is a high-

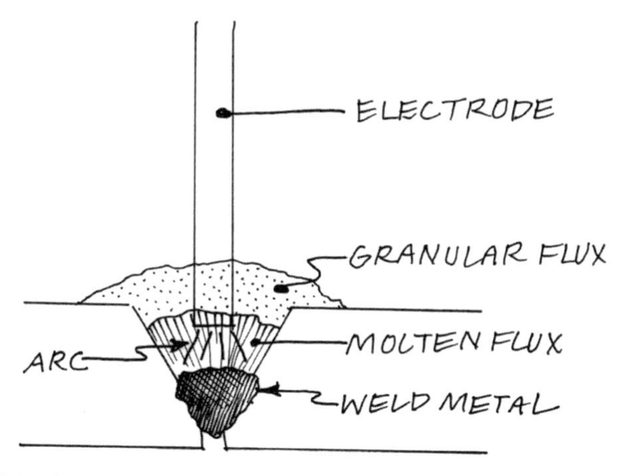

Figure 8.6 Submerged arc weld

deposition, automatic welding process for welding thick sections, Figure 8.5.

- Gas–Metal Arc Welding (GMAW)—consumes a continuous wire electrode protected by shielding gas only, shown in Figure 8.7. Some wires are tubular and contain additional alloying elements to improve operating and metallurgical characteristics. This process is very versatile and commonly used for welded fabrication.

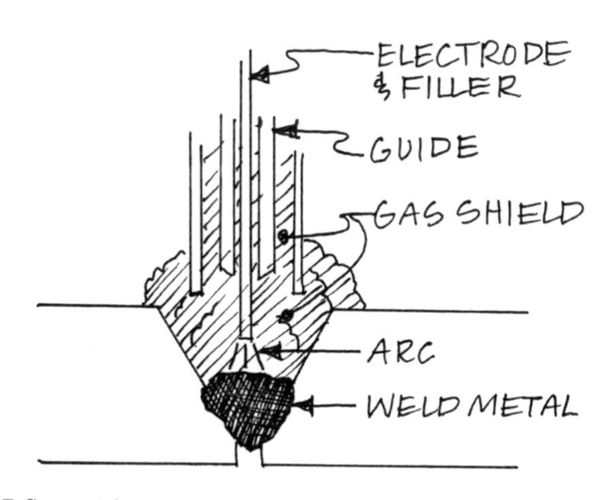

Figure 8.7 Gas–metal arc weld

- Flux-Cored Arc Welding (FCAW)—consists of a tubular, continuous wire electrode. Flux is formed into the tube creating shielding gas and slag in the welding arc, illustrated in Figure 8.8. It is like a "stick" electrode turned inside out. Some variations incorporate a secondary shielding gas for precision arc control and increased toughness.
- Electroslag Arc Welding (ESG)—uses highly specialized welding equipment and copper cooling "shoes", schematically shown in Figure 8.9. The process deposits large amounts of weld metal, joining base metals as thick as 4 in (100 mm) in a single pass. The cooling shoes pull the massive amount of heat out of the connection, mold the weld bead, and maintain the dimensional integrity of the connection.

Weldability of materials is fundamentally important to high-quality welds. Common structural steels, such as A36, A572, and A992 are all weldable. However, historic steels, high-strength low alloy, and reinforcing steels require more consideration. An effective way to estimate weldability of steel is to determine the **carbon equivalent** *CE* of the alloy. There are several equations used to determine the CE of a steel, depending upon the product form. Steel mills calculate the CE of common structural shapes and plate using Equation 8.1.

$$CE = C + \left(\frac{Mn + Si}{6}\right) + \left(\frac{Cr + Mo + V}{5}\right) + \left(\frac{Ni + Cu}{15}\right) \tag{8.1}$$

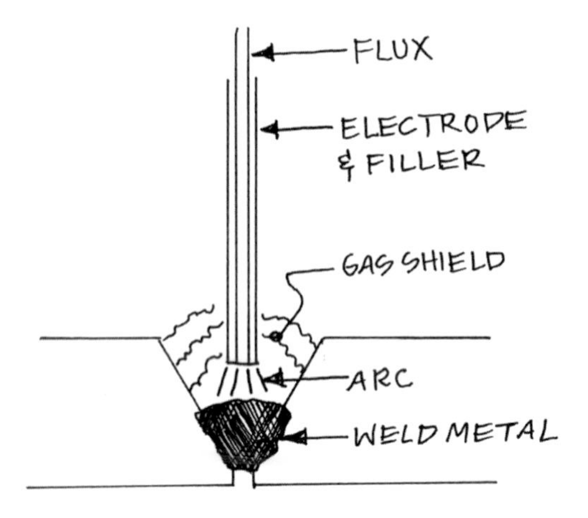

Figure 8.8 Flux cored arc weld

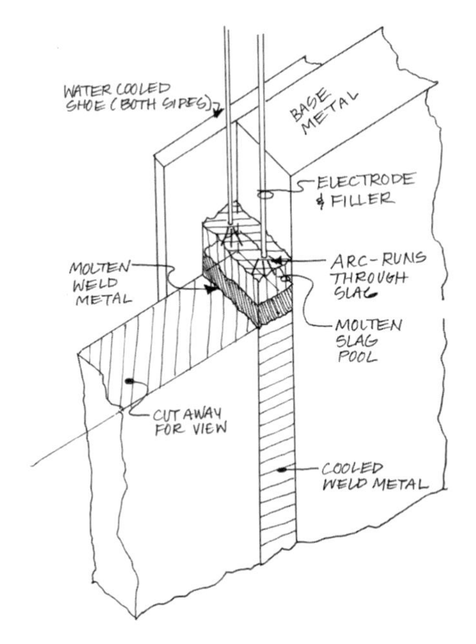

Figure 8.9 Electroslag arc weld

The *CE* and actual chemical content are printed on the material test reports (MTRs) from the mill. Carbon equivalence of reinforcing steels is calculated using modifications of Equation 8.1.

If the *CE* is less than 0.4, the steel has very good weldability. As it increases, low-hydrogen filler metals will be required to prevent cracking. If the *CE* is even higher, preheat is required. Preheat not only depends upon the steel chemistry, but upon section thickness and initial base metal temperature. Preheating the steel slows down the rate of weld cooling to develop optimum steel grain structures and tough weldments.

Joint design is also fundamental to sound, economical welds. Construction codes will determine what joint designs are appropriate for a given load path. The best weld joint develops the required capacity with the least amount of weld metal and labor. Overwelding wastes weld metal, increases labor hours, causes distortion, and leaves higher residual stresses. Fillet welds require the least preparation but suffer a design penalty because of the inherent stress-riser at the unwelded portion of the joint. **Complete joint penetration** welds (CJP), like Figure 8.10, develop the full strength of the section but require the most preparation and weld time. **Partial joint penetration** welds (PJP) reduce preparation and welding time, but suffer the same design penalty for the unwelded portion of the joint. Common joint types and preparation are shown in Figure 8.11.

Welds are specified on drawings using welding symbols. The engineer of record specifies the weld size, length, and general joint preparation. It is best to not over specify the joint preparation. The steel detailer will then add more information to weld symbol and piece cuts, once the fabrication shop is chosen. Figure 8.12 shows common welding symbols and their meaning. Consult AISC[7] for an expanded figure. Figure 8.13 shows example applications of these symbols.

Quality is of great consideration for welded joints, and the source of much effort during fabrication and erection. Common weld discontinuities

Figure 8.10 Completed CJP weld, with runoff tabs, shop access Courtesy S&S Steel fabrication

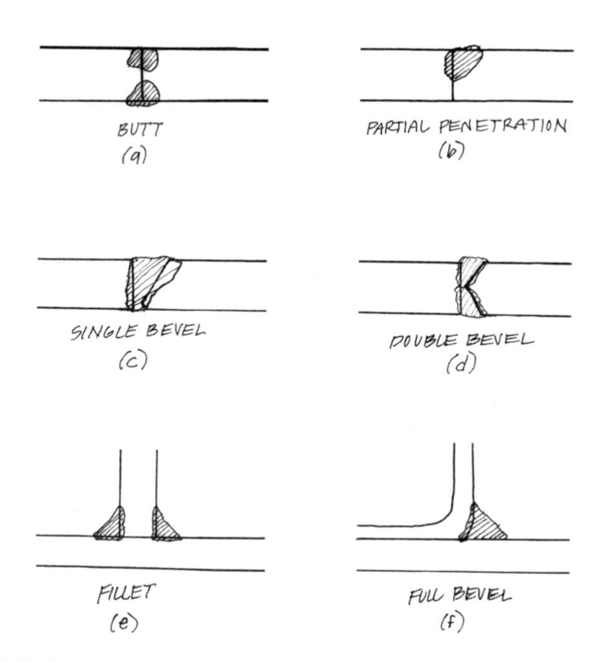

Figure 8.11 Common weld joint and preparation types

are incomplete fusion, lack of penetration, porosity, slag inclusions, undercutting, and cracking, illustrated in Figure 8.14. Weld inspection methods include visual, ultrasonic, magnetic particle (Figure 8.15), and radiographic examination. Weld inspection is extensively discussed in Chapter 10 of *Special Structural Topics* in this series.

You may be asked whether to specify bolting or welding on a project. The answer is yes. Yet there are times that one is preferred over the other. Strength may have little to do with the final decisions. Table 8.1 lists some advantages and disadvantages of each joining method.

8.1.3 Connected Elements

Connected elements are the connected portions of members and plate and angle used to make the connection. Figure 8.16 illustrates these in a column splice. In this example, we pay attention to the design of the plate, and flange and web with bolt holes.

WELD SYMBOL DESCRIPTION

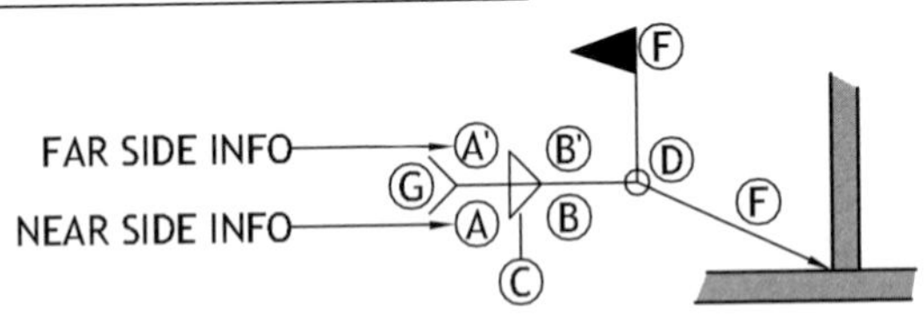

(A) Weld Size, always to left of symbol

(B) Weld Length or Pattern

(A') Weld Size (far side), leave blank if same size

(B') Weld Length or Pattern (far side), leave blank is same

(C) Weld Symbol

(D) Weld all around symbol

(E) Field weld symbol

(F) Weld leader, arrow side is bottom of weld information

(G) Notes

WELD TYPES

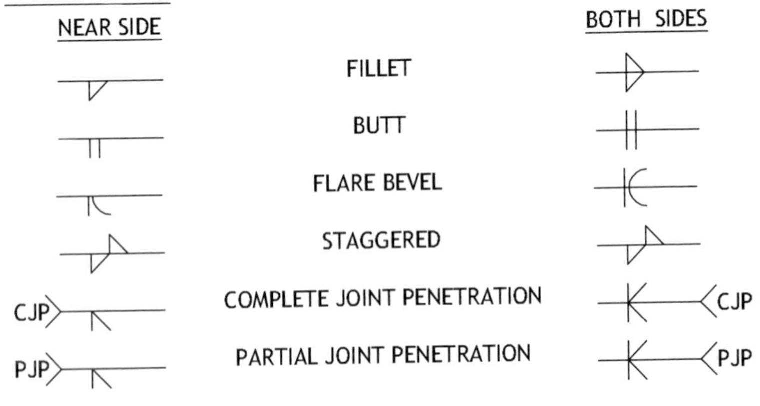

Figure 8.12 Weld symbol parts and definitions

JOINT AS DRAWN | JOINT AS WELDED

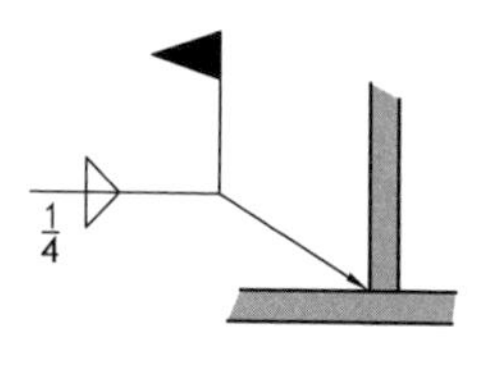

$\frac{1}{4}$ in fillet weld on both sides of joint, continuous in length

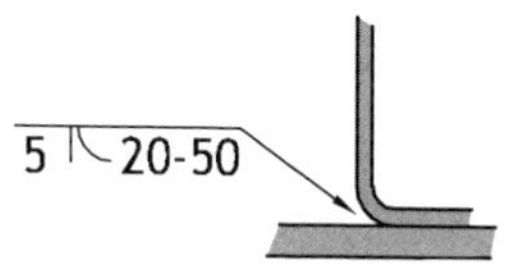

5 mm flare bevel weld on the near side of the joint. 20 mm long every 50 mm.

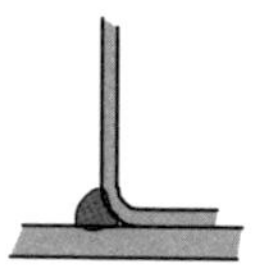

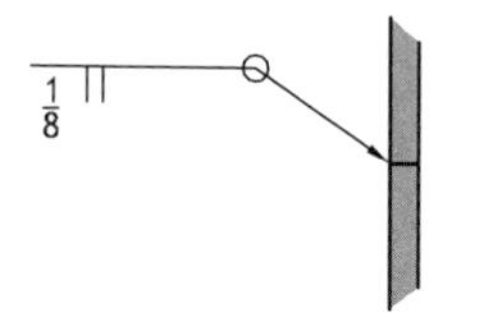

$\frac{1}{8}$ in butt weld, all around.

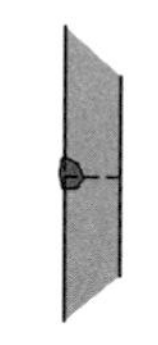

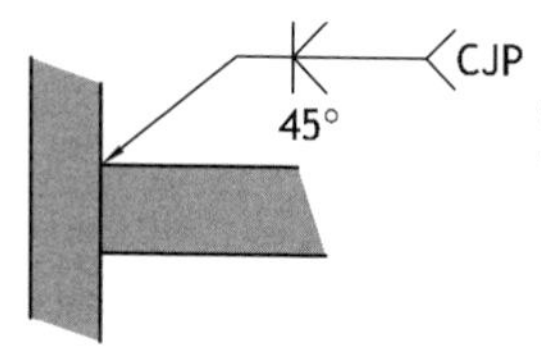

Complete joint penetration weld, continuous in length.

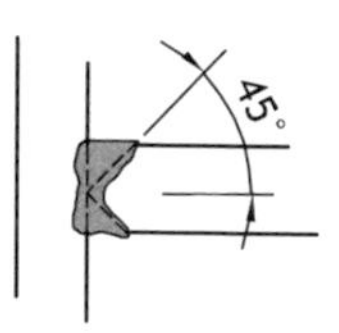

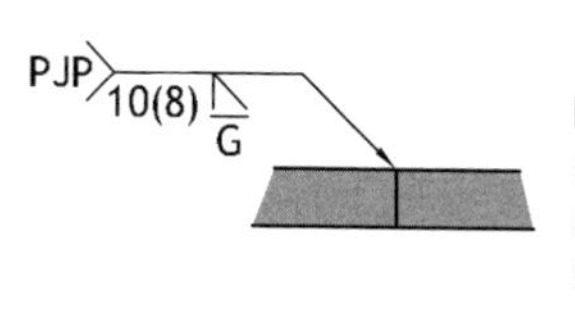

Partial penetration weld, continuous in length. 10 mm joint prep, and 8 mm minimum weld throat. Arrow side ground flat.

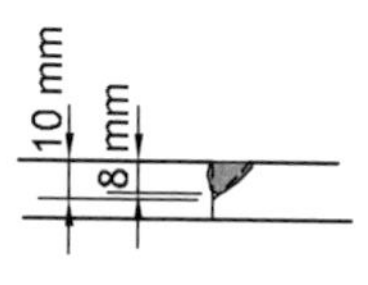

Figure 8.13 Weld symbol examples

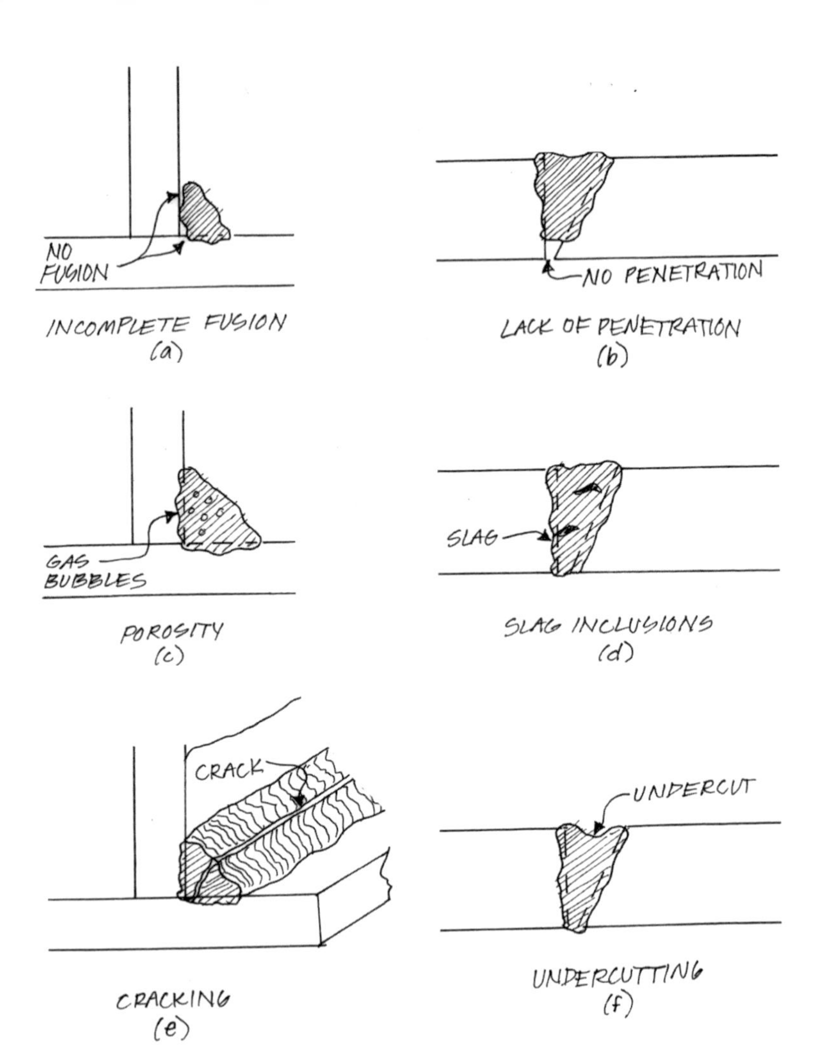

Figure 8.14 Weld discontinuities

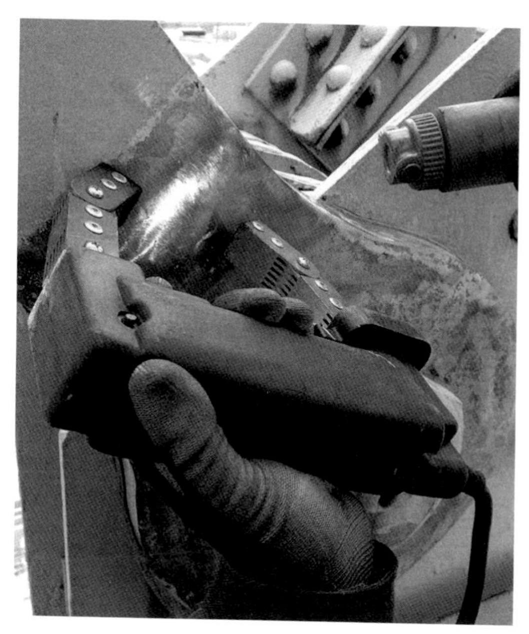

Figure 8.15 Magnetic particle testing

Table 8.1 Advantages and disadvantages of bolting and welding

Bolting	
Advantages	*Disadvantages*
Require less inspection	Reduces cross section
Field installation requires less skill	Larger joints
Parts can be easily disassembled	Require access to both sides of joint
Field installation requires less skill	Can become loose in vibrating conditions
Welding	
Advantages	*Disadvantages*
Potentially smaller joints	Require greater inspection
Join parts with unusual geometry	Induces residual stresses into joint
Alterations are easier	Field welding requires greater care

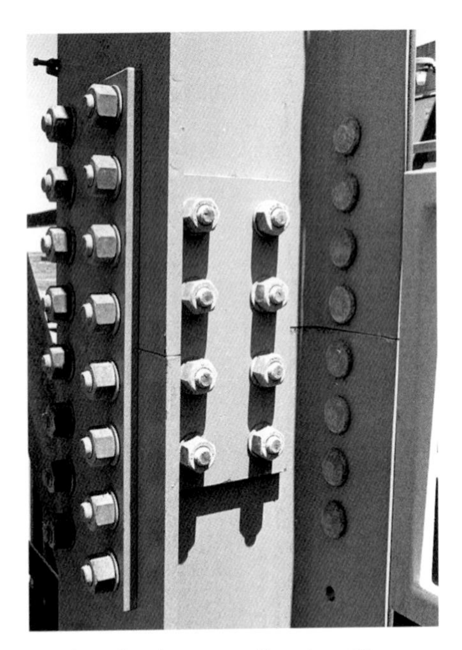

Figure 8.16 Column splice showing connecting elements

8.2 CAPACITY

8.2.1 Bolts

Bolts fail in tension, shear, bearing, or tearout in the base material, illustrated in Figure 8.17. These modes can exist for either tension or shear loads, depending on how we configure the joint. Remembering the chain analogy in Figure 3.5, we work our way through each member of a joint, designing each link in the load path.

We find bolt strength in tension and shear from Equation 8.2, denoting nominal capacity with R_n:

$$R_n = F_n A_b \tag{8.2}$$

where

F_n = nominal tensile stress F_{nt}, or shear stress F_{nv}, lb/in^2 (MN/m^2)
A_b = unthreaded area of the bolt, in^2 (m^2)
ϕ = 0.75 for bolt strength.

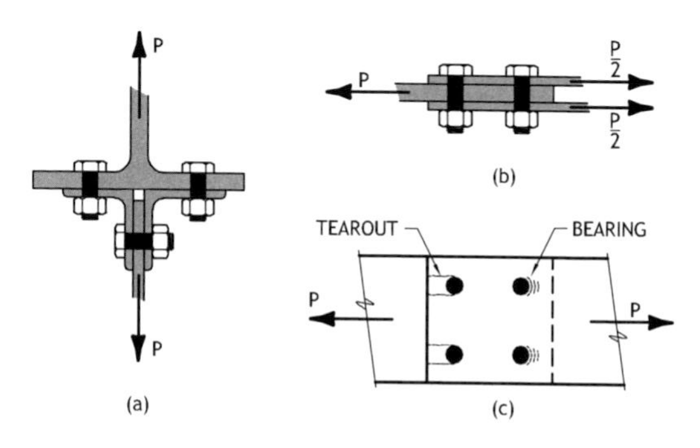

Figure 8.17 Bolt failure modes: (a) tension, (b) shear, (c) bearing and tearout

Table 8.2 Nominal bolt stresses

Fastener Type	*Nominal Strength, lb/in² (MN/m²)*		
	Tension	*Shear*[1]	
		Included	*Excluded*
A307	45	27	27
	(310)	(188)	(188)
Group A (A325)	90	54	68
	(620)	(372)	(469)
Group B (A490)	113	68	84
	(780)	(469)	(579)
Other[2]	$0.75F_u$	$0.45F_u$	$0.563F_u$

Source: AISC Steel Construction Manual, 14th Edition

Notes

1) Included, threads are in shear plane. Excluded, threads are not in shear plane.

2) Threaded parts meeting Section A3.4 of the *Manual*.

It may seem odd that we use the unthreaded bolt area. And it would be if the code didn't account for it in the nominal stress values listed in Table 8.2. These values are reduced for bolts in tension and shear through the threads.

When bolts are loaded in shear, we need to also look at bearing and tearout, where the bolt body bears on or tears through the connection material. The **nominal strength** is given by Equations 8.3 and 8.4. Take the lower.

$$R_n = 2.4dtF_u \text{ (bearing)} \quad (8.3)$$

$$R_n = 1.2l_c tF_u \text{ (tearout)} \quad (8.4)$$

where

l_c = clear distance between hole edges or hole edge and plate edge, in the direction of force, illustrated in Figure 8.18, in (mm):

t = material thickness, in (mm)

F_u = ultimate strength, lb/in^2 (MN/m^2)

D = bolt diameter, in (mm)

ϕ = 0.75 for bolt bearing and tearout.

The equation above assumes that hole deformation at service loads is a concern, which it often is. When it is not, we increase the equation by 25%.

Remember to apply the strength reduction factor ϕ, to get ϕR_n.

To speed up the design process, Tables 8.3, 8.4, and 8.5 present, respectively: bearing bolt shear strength, bolt tension strength, and bolt bearing strength.

Slip critical bolt strength is out of the scope of this book. However, their shear capacities are approximately 45% less than bearing-type bolts.

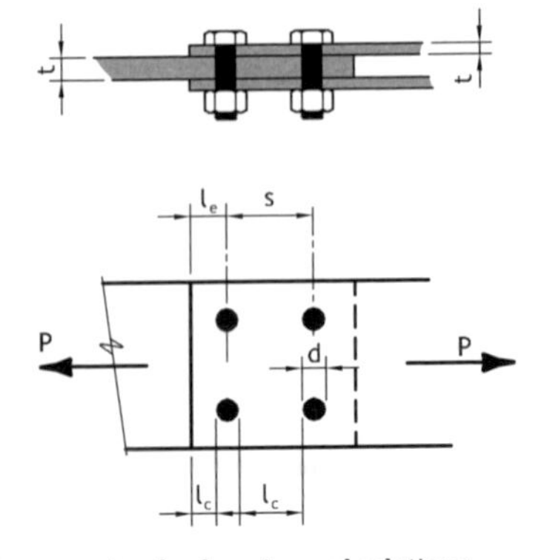

Figure 8.18 Bolt hole geometry for bearing calculations

Table 8.3 Bearing bolt shear strength

Group	*Bolt Grade*	*Imperial—Shear Strength, ϕr_n (k)* *Bolt Diameter (in)*							
		5/8	*3/4*	*7/8*	*1*	*1 1/8*	*1 1/4*	*1 3/8*	*1 1/2*
A	A325	12.4	17.9	24.3	31.8	40.3	49.8	59.9	71.7
B	A490	15.7	22.5	30.7	40.0	50.7	62.7	75.5	90.3
	A307	6.2	9.0	12.2	15.9	20.2	25.0	30.0	35.9

Group	*Bolt Grade*	*Metric—Shear Strength, ϕr_n (kN)* *Bolt Diameter (mm)*							
		16	*20*	*22*	*24*	*27*	*30*	*36*	
A	A325M	56.1	87.7	106.1	126	159.7	197.2	284	
B	A490M	70.7	111	134	159	201	249	358	
	A307	28.0	43.8	53.0	63.1	79.9	98.6	142	

Source: AISC Steel Construction Manual, 14th Edition

Note

1) Values are for single shear, with threads included in the shear plane. Higher values possible when threads are excluded.

Table 8.4 Bolt tension strength

Group	*Bolt Grade*	*Imperial—Tension Strength, ϕr_n (k)* *Bolt Diameter (in)*							
		5/8	*3/4*	*7/8*	*1*	*1 1/8*	*1 1/4*	*1 3/8*	*1 1/2*
A	A325	20.7	29.8	40.6	53.0	67.1	82.8	100	119
B	A490	26.0	37.4	51.0	66.6	84.2	104	126	150
	A307	10.4	14.9	20.3	26.5	33.5	41.4	50.1	59.6

Group	*Bolt Grade*	*Metric—Tension Strength, ϕr_n (kN)* *Bolt Diameter (mm)*							
		16	*20*	*22*	*24*	*27*	*30*	*36*	
A	A325	93.5	146.1	176.8	210	266	329	473	
B	A490	118	184	222	265	335	414	595	
	A307	46.7	73.0	88.4	105.2	133	164	237	

Source: AISC Steel Construction Manual, 14th Edition

Table 8.5 Bolt tension strength

End Distance	*Imperial—Bearing & Tearout Strength per inch Thickness* ϕr_n *(k/in)*							
	Bolt Diameter (in)							
l_e *(in)*	*5/8*	*3/4*	*7/8*	*1*	*1 1/8*	*1 1/4*	*1 3/8*	*1 1/2*
1	34.3	31.0						
1.25	47.3	44.0	40.8	37.5				
1.5	60.4	57.1	53.8	50.6	47.3			
1.75	65.3	70.1	66.9	63.6	60.4	57.1	53.8	
2	65.3	78.3	79.9	76.7	73.4	70.1	66.9	63.6
2.25	65.3	78.3	91.4	89.7	86.5	83.2	79.9	76.7
2.75	65.3	78.3	91.4	104	113	109	106.0	102.8
Spacing	*Bolt Diameter (in)*							
s (in)	*5/8*	*3/4*	*7/8*	*1*	*1 1/8*	*1 1/4*	*1 3/8*	*1 1/2*
1.75	55.5							
2.00	65.3	62.0						
2.50	65.3	78.3	81.6					
3.00	65.3	78.3	91.4	101	94.6			
3.50	65.3	78.3	91.4	104	117	114	108	
4.00	65.3	78.3	91.4	104	117	131	134	127
4.50	65.3	78.3	91.4	104	117	131	144	153

Notes

1) Values are for standard holes. See AISC Steel Construction Manual for additional hole types
2) Values are for F_u=58 k/in^2 (400 MN/m^2)
3) Blank values indicates spacing does not meet code edge and spacing minimum
4) Values above dark line are controlled by tearout
5) Multiply values by plate thickness to get bearing capacity

Table 8.5m Bolt tension strength

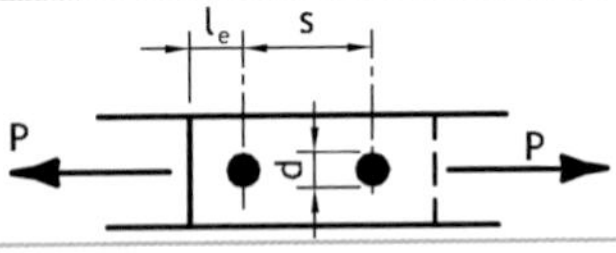

End Distance	*Metric—Bearing & Tearout Strength per mm Thickness* ϕr_n *(kN/mm)*						
	Bolt Diameter (mm)						
l_e *(mm)*	*16*	*20*	*22*	*24*	*27*	*30*	*36*
25	5.8						
30	7.6	6.8	6.5	5.9			
35	9.4	8.6	8.3	7.7	7.2		
40	11.2	10.4	10.1	9.5	9.0	8.5	
45	11.5	12.2	11.9	11.3	10.8	10.3	
50	11.5	14.0	13.7	13.1	12.6	12.1	11.0
60	11.5	14.4	15.8	16.7	16.2	15.7	14.6
70	11.5	14.4	15.8	17.3	19.4	19.3	18.2
Spacing	*Bolt Diameter (mm)*						
s (mm)	*16*	*20*	*22*	*24*	*27*	*30*	*36*
45	9.7						
50	11.5						
55	11.5	11.9					
65	11.5	14.4	14.8	13.7			
75	11.5	14.4	15.8	17.3	16.2		
90	11.5	14.4	15.8	17.3	19.4	20.5	
100	11.5	14.4	15.8	17.3	19.4	21.6	22.0
115	11.5	14.4	15.8	17.3	19.4	21.6	25.9

Notes

1) Values are for standard holes. See AISC Steel Construction Manual for additional hole types
2) Values are for F_u=58 k/in^2 (400 MN/m^2)
3) Blank values indicates spacing does not meet code edge and spacing minimum
4) Values above dark line are controlled by tearout
5) Multiply values by plate thickness to get bearing capacity

Prying is an important consideration for bolts in tension. Where the connected material is thin enough, it will deform and put a bending moment on the bolt body, shown in Figure 8.19. The easiest way to handle prying is to make the plate thick enough so it does not bend the bolt. We do this with Equation 8.5:

$$t_{min} = \sqrt{\frac{4T_u b'}{\phi p F_u}} \tag{8.5}$$

where

T_u = tension force in the bolt, k (kN)
b' = material cantilever, in (mm), defined in Figure 8.19.
ϕ = prying strength reduction factor
p = tributary length to the fastener, in (mm)
F_u = ultimate strength, lb/in^2 (MN/m^2).

Where it is not possible to thicken the material, AISC provides additional equations to determine the additional tension force from prying.

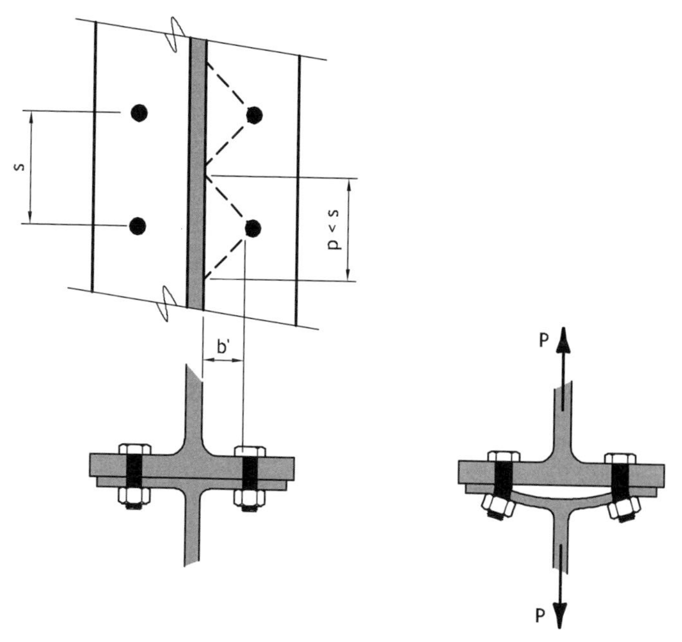

Figure 8.19 Prying in a bolted hanger connection

8.2.2 Welds

Weld strength is a function of the filler metal, base metal, and joint type. For complete joint penetration welds, the strength is determined by the base metal—discussed in the next section—providing the weld uses matching filler metal. For partial joint penetration, fillet, and **plug welds**, the base equation for capacity is shown by Equation 8.6:

$$R_n = F_{nw} A_{we} \tag{8.6}$$

where

F_{nw} = nominal weld metal strength, lb/in² (MN/m²)
= 0.6 F_{EXX} for tension and shear joints
= $0.9F_{EXX}$ for compression joints

F_{EXX} = filler metal classification strength, lb/in² (MN/m²)

A_{we} = effective weld area, in² (m²)
= al_w, shown in Figure 8.20.

ϕ = 0.75 for welds in shear, and 0.80 for partial penetration welds in tension.

The filler classification strength F_{EXX} is typically 60, 70, or 80 k/in² (414, 483, 552 MN/m²), with the middle value being very common.

To help with fillet weld sizing, Table 8.6 provides weld strength per inch (mm) for three filler metal strengths. We can multiply these values by weld length to get capacity. Or, we can divide a concentric joint load by the weld unit strength, to get the required length. You can also use it to check your homework.

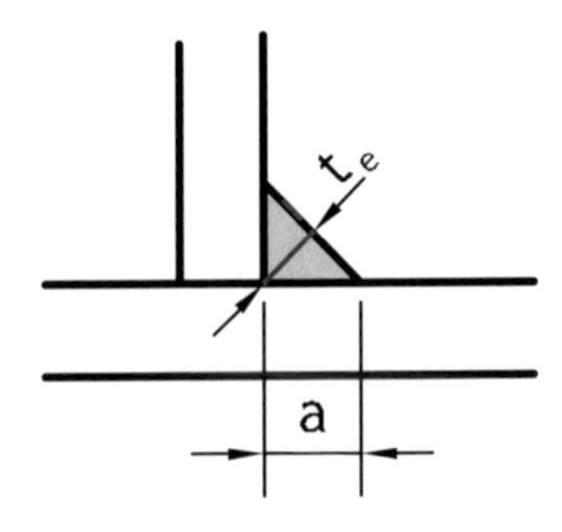

Figure 8.20 Weld throat determination

Table 8.6 Fillet weld capacity

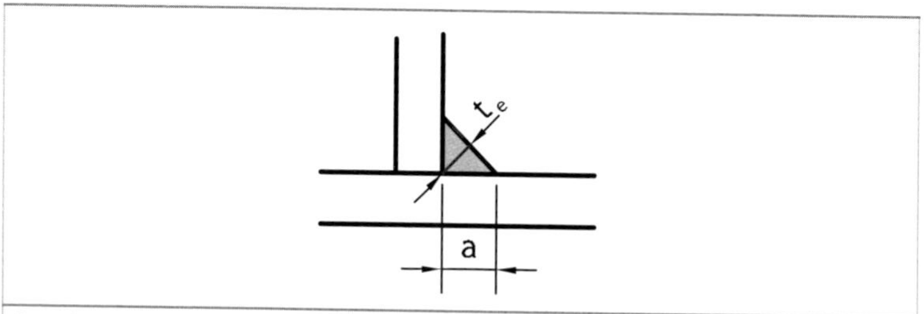

Imperial Measures				
Leg	*Throat*	*Weld Strength per inch, ϕr_n (k/in)*		
a	t_e	*Filler Strength, F_{EXX} (k/in²)*		
(in)	*(in)*	*60*	***70***	*80*
1/8	0.1	2.39	**2.78**	3.18
3/16	0.1	3.58	**4.18**	4.77
1/4	0.2	4.77	**5.57**	6.36
5/16	0.2	5.97	**6.96**	7.95
3/8	0.3	7.16	**8.35**	9.54
7/16	0.3	8.35	**9.74**	11.14
1/2	0.4	9.54	**11.1**	12.7
9/16	0.4	10.7	**12.5**	14.3
5/8	0.4	11.9	**13.9**	15.9
11/16	0.5	13.1	**15.3**	17.5
3/4	0.5	14.3	**16.7**	19.1

Note

1) See Table 8.12 for limits on fillet weld size

Table 8.6m Fillet weld capacity

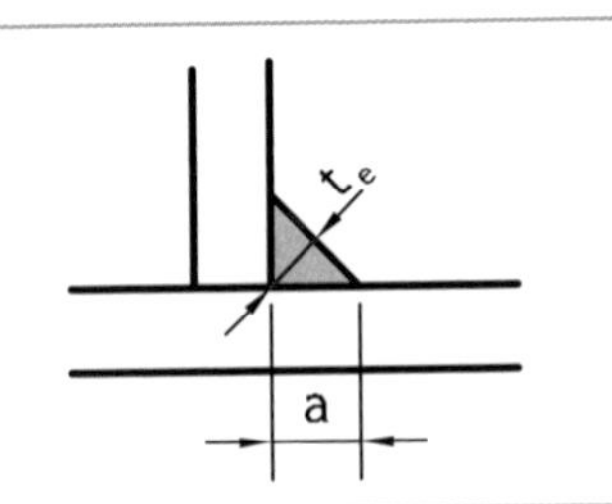

Metric Measures				
Leg	*Throat*	*Weld Strength per mm, ϕr_n (kN/mm)*		
a	t_e	*Filler Strength, F_{EXX} (MN/m^2)*		
(mm)	*(mm)*	*414*	***483***	*552*
3	2.1	0.395	**0.461**	0.527
4	2.8	0.527	**0.615**	0.702
5	3.5	0.659	**0.768**	0.878
6	4.2	0.790	**0.922**	1.05
7	4.9	0.922	**1.08**	1.23
8	5.7	1.05	**1.23**	1.40
10	7.1	1.32	**1.54**	1.76
12	8.5	1.58	**1.84**	2.11
14	9.9	1.84	**2.15**	2.46
16	11.3	2.11	**2.46**	2.81
18	12.7	2.37	**2.77**	3.16

Note

1) See Table 8.12 for limits on fillet weld size

8.2.3 Connected Elements

Connected elements have failure modes that parallel members; tension, shear, and compression. However, we add a fourth, block shear. We will discuss each of these modes in the paragraphs that follow:

For tension, we follow the equations in Sections 3.2.1 and 3.2.2. To be technically correct, we substitute P_n with R_n, to remind us we are designing connections.

For bending, follow the provisions of Chapter 4. However, flexure is an inefficient method to transfer connection forces and should be avoided when possible.

For shear, we parallel the direct shear method in Section 5.2.1, with the following slight modifications for strength (Equations 8.7 and 8.8):

$$R_n = 0.6F_yA_{gv} \text{ for shear yielding} \quad (8.7)$$

$$R_n = 0.6F_uA_{nv} \text{ for shear rupture} \quad (8.8)$$

where

F_y = tension yield strength, lb/in² (MN/m²)
F_u = tension ultimate strength, lb/in² (MN/m²)
A_{gv} = gross shear area, in² (m²)
A_{nv} = net shear area, in² (m²)
ϕ = 1.0 for shear yield, and 0.75 for shear rupture.

For compression, if $L_c/r \leq 25$, we don't need to worry about buckling and can use Equation 8.9:

$$R_n = F_yA_g \quad (8.9)$$

where

F_y = yield strength, lb/in² (MN/m²)
A_g = gross compression area, in² (m²)
ϕ = 0.90 for compression.

When $L_c/r > 25$, we use the buckling equations in Chapter 6.

Block shear occurs where a portion of a connected element tears out in a combination of shear and tension, illustrated in Figure 8.21. We determine block shear capacity as the lesser of the following two equations (8.10). The first checks shear and tension rupture, the second checks shear yield and tension rupture.

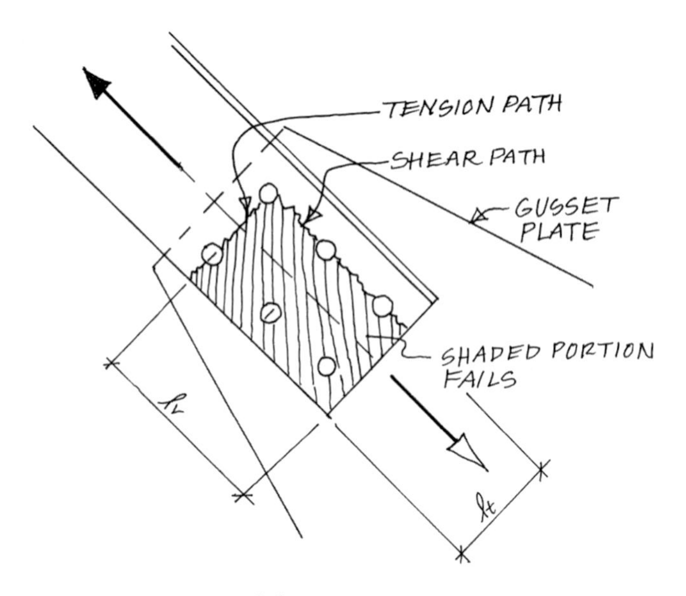

Figure 8.21 Block shear in angle brace

$$R_n = 0.6F_uA_{nv} + U_{bs}F_uA_{nt} \tag{8.10}$$
$$R_n = 0.6F_yA_{gv} + U_{bs}F_uA_{nt}$$

where

F_u = tension ultimate strength, lb/in^2 (MN/m^2)
F_y = tension yield strength, lb/in^2 (MN/m^2)
A_{nv} = net shear area, in^2 (m^2)
A_{gv} = gross shear area, in^2 (m^2)
A_{nt} = net tension area, in^2 (m^2)
U_{bs} = shear lag factor,
= 1.0 for uniform tension stress, 0.5 for non-uniform stress
ϕ = 0.75 for block shear.

8.3 DEMAND VS CAPACITY

Once we have the demand (from structural analysis) and capacity, we compare the two, just like we did for member failure modes. When the capacity ϕR_n, is greater than demand T_u or P_u, we know our connection has adequate strength. If not, we increase the number of bolts, length of weld, or thickness of plate and angle. Alternatively, we can reconfigure the joint, add more connectors, or change the structural geometry.

8.4 DETAILING

Connections have numerous detailing requirements to ensure they perform the way we intend. This section provides the basic detailing requirements for bolts and welds. To get started, the following figures are examples of common structural connections. See if you can figure out how the requirements in this section apply to the photos in Figures 8.22–8.25.

8.4.1 Bolts

Tables 8.7–8.11 provide dimensions and detailing requirements for bolted joints.

Figure 8.22 Heavy base plate and shear lug in shop

Shop access provided by S&S Steel

Figure 8.23 Double angle shear connection

Figure 8.24 Welded pipe brace

Courtesy Mark Steel

Figure 8.25 Bolted truss splice and web connection

Table 8.7 Bolt dimensions

	Measurement	*Fastener Dimensions (in)* *Nominal Bolt Diameter d (in)*								
		1/2	*5/8*	*3/4*	*7/8*	*1*	*1 1/8*	*1 1/4*	*1 3/8*	*1 1/2*
A325 and A490 Bolts	Width Across Flats F	7/8	1 1/16	1 1/4	1 7/16	1 5/8	1 13/16	2	2 3/16	2 3/8
	Height H	5/16	25/64	15/32	35/64	39/64	11/16	25/32	27/32	15/16
	Thread Length	1	1 1/4	1 3/8	1 1/2	1 3/4	2	2	2 1/4	2 1/4
	Bolt Length = Grip + Washer Thickness + →	11/16	7/8	1	1 1/8	1 1/4	1 1/2	1 5/8	1 3/4	1 7/8
A563 Nuts	Width Across Flats F	7/8	1 1/16	1 1/4	1 7/16	1 5/8	1 13/16	2	2 3/16	2 3/8
	Height H	31/64	39/64	47/64	55/64	63/64	1 7/64	1 7/32	1 11/32	1 15/32
F436 Circular Washers	Outside Diameter OD	1 1/16	1 5/16	1 15/32	1 3/4	2	2 1/4	2 1/2	2 3/4	3
	Inside Diameter ID	17/32	11/16	13/16	15/16	1 1/8	1 1/4	1 3/8	1 1/2	1 5/8
	Thickness T	0.097	0.122	0.122	0.136	0.136	0.136	0.136	0.136	0.136
	Min. Edge Distance E	7/16	9/16	21/32	25/32	7/8	1	1 3/32	1 7/32	1 5/16
F436 Tapered Washers	Width A	1 3/4	1 3/4	1 3/4	1 3/4	1 3/4	2 1/4	2 1/4	2 1/4	2 1/4
	Mean Thickness T	5/16	5/16	5/16	5/16	5/16	5/16	5/16	5/16	5/16
	Taper Slope	02:12	02:12	02:12	02:12	02:12	02:12	02:12	02:12	02:12
	Minimum Edge Distance E	7/16	9/16	21/32	25/32	7/8	1	1 3/32	1 7/32	1 5/16

Source: AISC Steel Construction Manual, 14th Edition

Table 8.7m Bolt dimensions

	Measurement	*Fastener Dimensions (mm)* *Nominal Bolt Diameter d (mm)*							
		M12	*M16*	*M20*	*M22*	*M24*	*M27*	*M30*	*M36*
A325 and A490 Bolts	Width Across Flats *F*	21.00	27.00	34.00	36.00	41.00	46.00	50.00	60.00
	Height *H*	7.95	10.75	13.40	14.90	15.90	17.90	19.75	23.55
	Thread Length	25.0	31.0	36.0	38.0	41.0	44.0	49.0	56.0
	Bolt Length = Grip + Washer Thickness + →	18.0	22.0	26.0	29.0	32.0	38.0	42.0	45.0
A563 Nuts	Width Across Flats *F*	21.0	27.0	34.0	36.0	41.0	46.0	50.0	60.0
	Height *H*	12.3	17.1	20.7	23.6	24.2	27.6	30.7	36.6
F436 Circular Washers	Outside Diameter *OD*	25.0	32.0	39.0	39.0	44.0	50.0	56.0	66.0
	Inside Diameter *ID*	13.5	17.5	22.0	23.0	26.0	28.0	33.0	39.0
	Thickness *T*	2.5	3.5	4.0	4.0	4.6	5.0	5.1	5.6
	Min. Edge Distance *E*	10.5	14.0	17.5	19.2	21.0	23.6	26.2	31.5
F436 Tapered Washers	Width *A*	30	36	44	50	56	56	62	68
	Mean Thickness *T*	4.1	5.0	6.1	6.5	6.9	6.9	7.5	8.0
	Taper Slope	14%	14%	14%	14%	14%	14%	14%	14%
	Minimum Edge Distance *E*	11.1	14.3	16.7	19.8	22.2	25.4	27.8	31.0

Note 1) M12 indicates metric bolt with 12 mm diameter

Table 8.8 Bolt areas

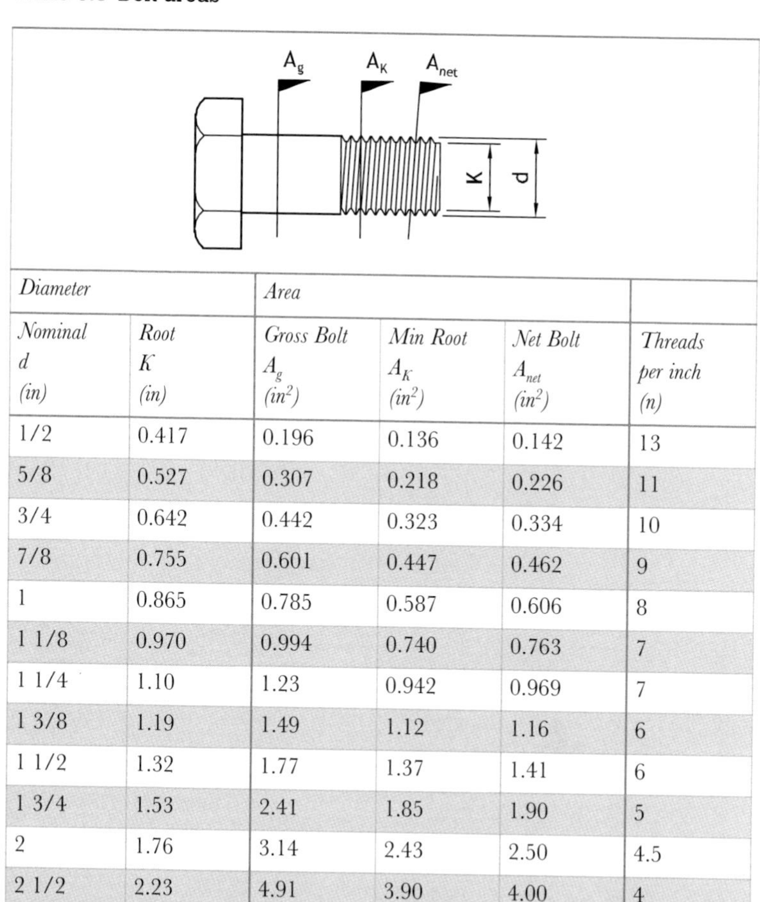

Diameter		*Area*			
Nominal d (in)	*Root K (in)*	*Gross Bolt A_g (in^2)*	*Min Root A_K (in^2)*	*Net Bolt A_{net} (in^2)*	*Threads per inch (n)*
1/2	0.417	0.196	0.136	0.142	13
5/8	0.527	0.307	0.218	0.226	11
3/4	0.642	0.442	0.323	0.334	10
7/8	0.755	0.601	0.447	0.462	9
1	0.865	0.785	0.587	0.606	8
1 1/8	0.970	0.994	0.740	0.763	7
1 1/4	1.10	1.23	0.942	0.969	7
1 3/8	1.19	1.49	1.12	1.16	6
1 1/2	1.32	1.77	1.37	1.41	6
1 3/4	1.53	2.41	1.85	1.90	5
2	1.76	3.14	2.43	2.50	4.5
2 1/2	2.23	4.91	3.90	4.00	4
3	2.73	7.07	5.85	5.97	4
3 1/2	3.23	9.62	8.19	8.33	4
4	3.73	12.6	10.9	11.1	4

Source: AISC Steel Construction Manual, 14th Edition

Table 8.8m **Bolt areas**

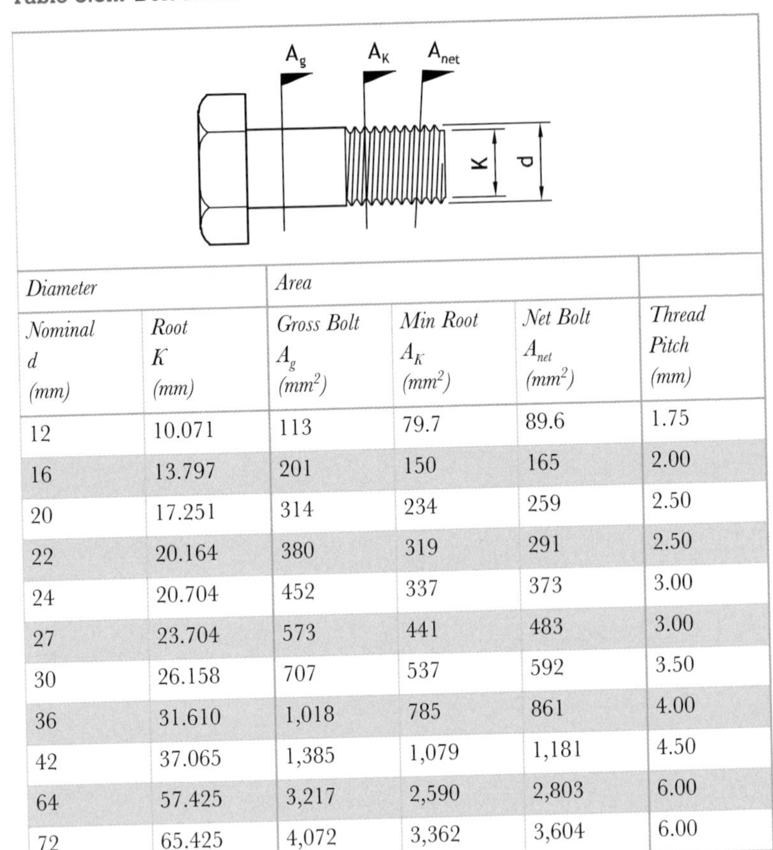

Diameter		*Area*			
Nominal d *(mm)*	*Root* K *(mm)*	*Gross Bolt* A_g *(mm²)*	*Min Root* A_K *(mm²)*	*Net Bolt* A_{net} *(mm²)*	*Thread Pitch (mm)*
12	10.071	113	79.7	89.6	1.75
16	13.797	201	150	165	2.00
20	17.251	314	234	259	2.50
22	20.164	380	319	291	2.50
24	20.704	452	337	373	3.00
27	23.704	573	441	483	3.00
30	26.158	707	537	592	3.50
36	31.610	1,018	785	861	4.00
42	37.065	1,385	1,079	1,181	4.50
64	57.425	3,217	2,590	2,803	6.00
72	65.425	4,072	3,362	3,604	6.00
80	73.425	5,027	4,234	4,506	6.00
90	83.425	6,362	5,466	5,774	6.00
100	93.425	7,854	6,855	7,197	6.00

Source: AISC Steel Construction Manual, 14th Edition

Table 8.9 Tightening clearances

Imperial Measures

Bolt Diameter	Socket Diameter	H_1	H_2	C_1	C_2	C_3 (in)	
(in)	(in)	(in)	(in)	(in)	(in)	Circular	Clipped
5/8	1 3/4	25/64	1 1/4	1	11/16	11/16	9/16
3/4	2 1/4	15/32	1 3/8	1 1/4	3/4	3/4	11/16
7/8	2 1/2	35/64	1 1/2	1 3/8	7/8	7/8	13/16
1	2 5/8	39/64	1 5/8	1 7/16	15/16	1	7/8
1 1/8	2 7/8	11/16	1 7/8	1 9/16	1 1/16	1 1/8	1
1 1/4	3 1/8	25/32	2	1 11/16	1 1/8	1 1/4	1 1/8
1 3/8	3 1/4	27/32	2 1/8	1 3/4	1 1/4	1 3/8	1 1/4
1 1/2	3 1/2	15/16	2 1/4	1 7/8	1 5/16	1 1/2	1 5/16

Metric Measures

Bolt Diameter	Socket Diameter	H_1	H_2	C_1	C_2	C_3 (mm)	
(mm)	(mm)	(mm)	(mm)	(mm)	(mm)	Circular	Clipped
M16	44	10	32	25	17	17	14
M20	57	12	35	32	19	19	17
M22	64	14	38	35	22	22	21
M24	67	15	41	37	24	25	22
M27	73	17	48	40	27	29	25
M30	79	20	51	43	29	32	29
M36	89	24	57	48	33	38	33

Note

1) For bolts aligned in a row. See the *Manual* for staggered bolts

Source: AISC Steel Construction Manual, 14th Edition

Table 8.10 Hole and slot standard sizes

Imperial Measures (in)				
	Hole Dimensions (in)			
Bolt Diameter	*Standard Diameter*	*Oversize Diameter*	*Short Slot Width × Length*	*Long Slot Width × Length*
1/2	9/16	5/8	9/16 × 11/16	9/16 × 1 1/4
5/8	11/16	13/16	11/16 × 7/8	11/16 × 1 9/16
3/4	13/16	15/16	13/16 × 1	13/16 × 1 7/8
7/8	15/16	1 1/16	15/16 × 1 1/8	15/16 × 2 3/16
1	1 1/16	1 1/4	1 1/16 × 1 5/16	1 1/16 × 2 1/2
≥1 1/8	d+1/16	d+5/16	(d+1/16) × (d+3/8)	(d+1/16) × (d+2.5d)

Metric Measures (mm)				
	Hole Dimensions (mm)			
Bolt Diameter	*Standard Diameter*	*Oversize Diameter*	*Short Slot Width × Length*	*Long Slot Width × Length*
M16	18	20	18 × 22	18 × 40
M20	22	24	22 × 26	22 × 50
M22	24	28	24 × 30	24 × 55
M24	27	30	27 × 32	27 × 60
M27	30	35	30 × 37	30 × 67
M30	33	38	33 × 40	33 × 75
≥M36	d+3	d+8	(d+3) × (d+10)	(d+3) × (d+2.5d)

Source: AISC Steel Construction Manual, 14th Edition

Table 8.11 Spacing and edge distance requirements

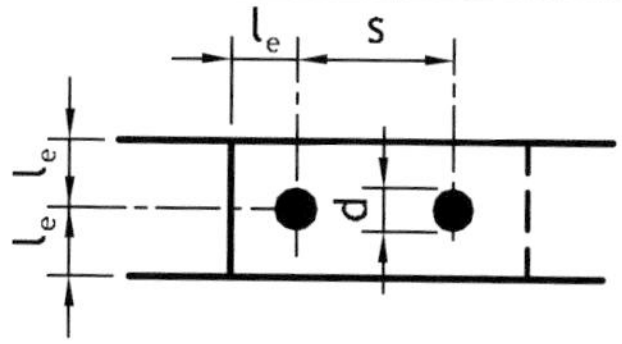

Imperial Measures (in)				
	Spacing		*Edge Distance*	
Bolt Diameter	*Preferred*[1]	*Minimum*	*Standard Hole*	*Oversize Hole*
1/2	2	1.33	3/4	13/16
5/8	3	1.67	7/8	15/16
3/4	3	2.00	1	1 1/16
7/8	3	2.33	1 1/8	1 3/16
1	3	2.67	1 1/4	1 3/8
1 1/8	3 3/8	3.00	1 1/2	1 11/16
1 1/4	3 3/4	3.33	1 5/8	1 13/16
≥ 1 1/4	3d	(2 2/3)d	1.25d	1.25d+3/16

Metric Measures (mm)				
Spacing	*Edge Distance*			
Bolt Diameter	*Preferred*[1]	*Minimum*	*Standard Hole*	*Oversize Hole*
16	50	43	22	24
20	75	54	26	28
22	75	59	28	30
24	75	64	30	33
27	80	72	34	37
30	90	80	38	41
34	100	91	46	49
≥ 36	3d	(2 2/3)d	1.25d	1.25d+3

Note

1) Based on AISC recommendations and common shop practice

Source: AISC Steel Construction Manual, 14th Edition

8.4.2 Welds

Welds have limits on their minimum and maximum sizes. Table 8.12 provides these for fillet welds. Partial penetration welds have a similar table in the *Steel Manual.*

8.5 DESIGN STEPS

Now that we have the basics, the following steps will guide you when designing a connection.

Step 1: Determine the force in the connector group and/or individual connector.

Step 2: Choose the fastener type. This decision is a function of load magnitude, constructability, contractor preference, and aesthetics.

Step 3: Configure the joint geometry.

Step 4: Find the connector strength by equation or from the applicable table in this chapter.

Step 5: Compare the strength to the connection load.

Step 6: Sketch the final connection geometry.

8.6 DESIGN EXAMPLE

In this example, we will assume the architect has decided that the 62 ft (18.9 m) beam in the example in Chapter 5 should be a truss to allow ductwork to pass between the webs.

Step 1: Determine Forces

From a truss analysis, we determined the axial tension force in the connection is

$T_u = 48 \text{ k}$	$T_u = 214 \text{ kN}$

See Chapter 10 of *Introduction to Structures*, in this series, for additional information on truss analysis.

Step 2: Choose Fastener Type

We will use bearing bolts for this connection.

Table 8.12 Fillet weld size limitations

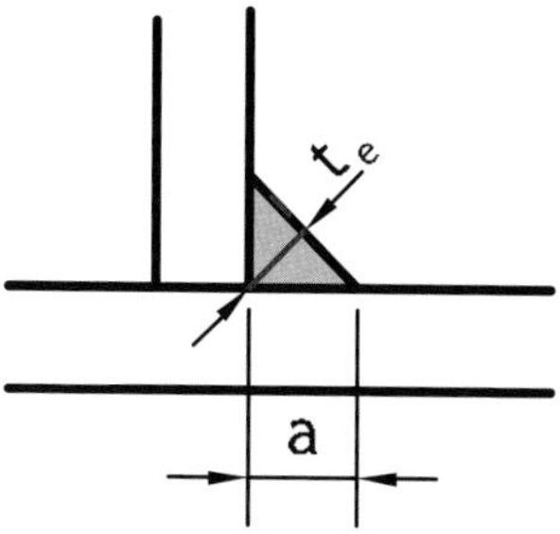

Minimum Fillet Weld Sizes			
Imperial Measures		*Metric Measures*	
Material Thickness (in)	*Minimum Weld Leg, a (in)*	*Material Thickness (mm)*	*Minimum Weld Leg, a (mm)*
1/4	1/8	6.0	3.0
Over 1/4 to 1/2	3/16	Over 6 to 13	5.0
Over 1/2 to 3/4	1/4	Over 13 to 19	6.0
Over 3/4	5/16	Over 19	8.0

Maximum Fillet Weld Sizes	
Imperial Measures	
Material Edge Thickness	*Maximum Weld Leg, a*
less than 1/4 in	Thickness of Material
1/4 in or greater	Thickness of Material- 1/16 in
Metric Measures	
Material Edge Thickness	*Maximum Weld Leg, a*
less than 6 mm	Thickness of Material
6 mm or greater	Thickness of Material- 2 mm

Source: AISC Steel Construction Manual, 14th Edition

Step 3: Configure the Joint Geometry

We sketch out the joint geometry, considering member type, bolt spacing, and edge distance. Figure 8.26a shows our initial joint and pertinent information.

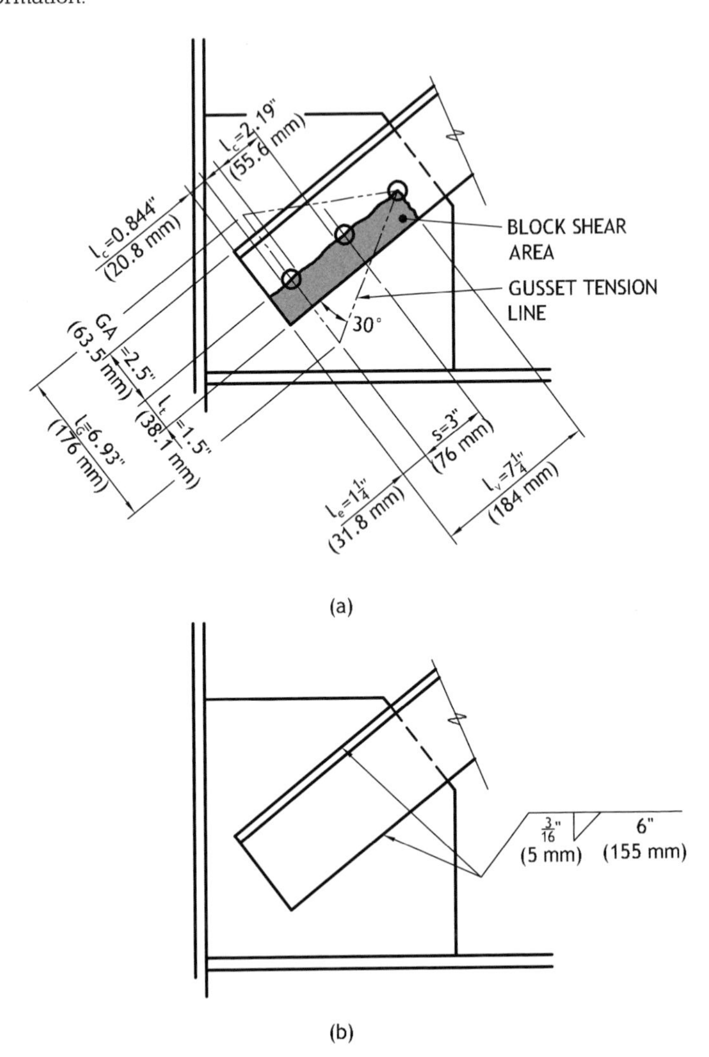

Figure 8.26 Example truss joint configuration

Step 4: Find Connector Strength

We have several failure modes to check. Starting at the member and moving to the gusset plate, we will look at:

- Net Section Rupture in the angle
- Bolt Bearing
- Bolt Shear
- Block Shear.

We don't need to check gross yield on the member, as we would have checked this when we sized it.

Step 4a: Net Section Rupture

From our sketch, we see we only have one bolt hole in the cross-section.

The base equation is:

$$R_{nR} = A_e F_u$$

To find effective area A_e we will need the following information on the member and bolt sizes.

We have selected a single L4 × 4 × 3/8 (L102 × 102 × 9.5), and 3/4 in (M20) bolt. This yields

$$t = 0.375 \text{ in}$$
$$d = 3/4 \text{ in}$$

$$t = 9.5 \text{ mm}$$
$$d = 20 \text{ mm}$$

$$\begin{aligned} d_{\text{hole}} &= d + 1/8 \\ &= 0.75 \text{ in} + 1.8 \text{ in} \\ &= 0.875 \text{ in} \end{aligned}$$

$$\begin{aligned} d_{\text{hole}} &= d + 2 \\ &= 20 \text{ mm} + 2 \text{ mm} \\ &= 22 \text{ mm} \end{aligned}$$

From Table A1.3 we find the gross area of the angle as

$$A_g = 2.86 \text{ in}^2$$

$$A_g = 1{,}850 \text{ mm}^2$$

The effective area is given by

$$\begin{aligned} A_e &= UA_{\text{net}} \\ &= U\left(A_g - A_{\text{hole}}\right) \end{aligned}$$

$$\begin{aligned} A_{\text{hole}} &= td_{\text{hole}} \\ &= 0.375 \text{ in}(0.875 \text{ in}) \\ &= 0.328 \text{ in}^2 \end{aligned}$$

$$\begin{aligned} &= 9.5 \text{ mm}(22 \text{ mm}) \\ &= 209 \text{ mm}^2 \end{aligned}$$

With three bolts, and using Table 3.1, $U = 0.6$

$$A_e = 0.6\left(2.86\text{ in}^2 - 0.328\text{ in}^2\right) = 1.52\text{ in}^2$$

$$A_e = 0.6\left(1{,}850\text{ mm}^2 - 209\text{ mm}^2\right) = 985\text{ mm}^2$$

Knowing

$$F_u = 58\text{ k/in}^2$$

$$F_u = 400\text{ MN/m}^2$$

we solve for the nominal rupture capacity, denoted by a subscript R.

$$\begin{aligned} R_{nR} &= A_e F_u \\ &= 1.52\text{ in}^2\left(58\text{ k/in}^2\right) \\ &= 88.2\text{ k} \end{aligned}$$

$$\begin{aligned} &= 985\text{ mm}^2\left(400\text{ MN/m}^2\right)\left(\frac{1\text{ m}}{1000\text{ mm}}\right)^2\left(\frac{1000\text{ kN}}{1\text{ MN}}\right) \\ &= 394\text{ kN} \end{aligned}$$

Multiplying by $\phi = 0.75$, we get

$$\begin{aligned} \phi R_{nR} &= 0.75 R_{nR} \\ &= 0.75(88.2\text{ k}) \\ &= 66.2\text{ k} \end{aligned}$$

$$\begin{aligned} &= 0.75(394\text{ kN}) \\ &= 296\text{ kN} \end{aligned}$$

Since this is greater than the demand T_u, we are OK.

Step 4b: Bolt Bearing

The load now moves from the brace into the interface with the bolt, known as bolt bearing. We take the lower of the following equations:

$$\begin{aligned} R_{nBl} &= 1.2 l_c t F_u \\ R_{nBd} &= 2.4 d t F_u \end{aligned}$$

From our sketch, we see

$$\begin{aligned} l_c &= 1.25\text{ in} - \frac{13}{16}\Big/ 2\text{ in} \\ &= 0.844\text{ in} \end{aligned}$$

$$\begin{aligned} l_c &= 31.8\text{ mm} - \frac{22}{2}\text{ mm} \\ &= 20.8\text{ mm} \end{aligned}$$

Knowing the other inputs, we get

$$R_{nBl} = 1.2(0.844 \text{ in})0.375 \text{ in}\left(58 \text{ k/in}^2\right)$$
$$= 22.0 \text{ k}$$

$$R_{nBl} = 1.2(20.8 \text{ mm})9.5 \text{ mm}\left(0.4 \text{ kN/mm}^2\right)$$
$$= 94.9 \text{ kN}$$

Multiplying by $\phi = 0.75$, we get

$$\phi R_{nBl} = 0.75(22.0 \text{ k})$$
$$= 16.5 \text{ k}$$

$$\phi R_{nBl} = 0.75(94.9 \text{ kN})$$
$$= 71.2 \text{ kN}$$

Checking the bearing limit, based on diameter, we get

$$\phi R_{nBd} = 0.75(2.4dtF_u)$$
$$= 0.75\left[2.4(0.75 \text{ in})0.375 \text{ in}\left(58 \text{ k/in}^2\right)\right]$$
$$= 29.4 \text{ k}$$

$$= 0.75\left[2.4(20 \text{ mm})9.5 \text{ mm}\left(0.4 \text{ kN/mm}^2\right)\right]$$
$$= 137 \text{ kN}$$

The first equation is controlling. Based on this, we can determine how many bolts we need to use:

$$n = \frac{T_u}{\phi R_{nBl}}$$
$$= \frac{48 \text{ k}}{16.5 \text{ k/bolt}}$$
$$= 2.9 \text{ bolts}$$

$$= \frac{214 \text{ kN}}{71.2 \text{ kN/bolt}}$$
$$= 3.01 \text{ bolts}$$

We will use three bolts.

Step 4c: Bolt Shear

The force now causes a shear force through the body of the bolt. Assuming the threads are in the shear plane, we get the bolt capacity from Table 8.3.

$$\phi R_{nV} = 17.9 \text{ k}$$

$$\phi R_{nV} = 87.7 \text{ kN}$$

Since this is greater than the limiting bearing strength ϕR_{nBl}, we know the number of bolts we chose is sufficient.

Cool!

Step 4d: Block Shear

The last thing we need to check is block shear, where a piece of the angle could break out, shown in Figure 8.26. We take the minimum of the following two equations.

$$R_{nU} = 0.6F_uA_{nv} + U_{bs}F_uA_{nt}$$
$$R_{nY} = 0.6F_yA_{gv} + U_{bs}F_uA_{nt}$$

$$F_y = 36\ \text{k/in}^2 \qquad F_y = 248\ \text{MN/m}^2$$

Using Figure 8.26, we find the following lengths and areas.

$$l_v = 1.25\ \text{in} + 2(3\ \text{in}) = 7.25\ \text{in} \qquad l_v = 31.8\ \text{mm} + 2(76.2\ \text{mm}) = 184\ \text{mm}$$
$$l_t = 1.5\ \text{in} \qquad l_t = 38.1\ \text{mm}$$

$$A_{gv} = l_vt$$
$$= 7.25\ \text{in}(0.375\ \text{in}) \qquad = 184\ \text{mm}(9.5\ \text{mm})$$
$$= 2.72\ \text{in}^2 \qquad = 1{,}784\ \text{mm}^2$$

Knowing we have 2 1/2 holes reducing our area, our net shear area is

$$A_{nv} = A_{gv} - 2.5d_{\text{hole}}t$$
$$= 2.72\ \text{in}^2 - 2.5(0.875\ \text{in})0.375\ \text{in} \qquad = 1{,}748\ \text{mm}^2 - 2.5(22\ \text{mm})9.5\ \text{mm}$$
$$= 1.9\ \text{in}^2 \qquad = 1{,}225\ \text{mm}^2$$

Next, we get the net tension area:

$$A_{nt} = t\left(l_t - \frac{d_{\text{hole}}}{2}\right)$$
$$= 0.375\ \text{in}\left(1.5\ \text{in} - \frac{0.875\ \text{in}}{2}\right) \qquad = 9.5\ \text{mm}\left(38.1\ \text{mm} - \frac{22\ \text{mm}}{2}\right)$$
$$= 0.398\ \text{in}^2 \qquad = 258\ \text{mm}^2$$

We now calculate the two block shear equations, taking $U_{bs} = 1.0$, since we will have uniform stress in the tension portion

$$R_{nU} = 0.6F_uA_{nv} + U_{bs}F_uA_{nt}$$

$$R_{nU} = 0.6(58\ \text{k/in}^2)1.9\ \text{in}^2 + 1.0(58\ \text{k/in}^2)0.39\ \text{in}^2$$
$$= 89.2\ \text{k}$$

$$R_{nU} = 0.6(0.4\ \text{kN/mm}^2)1{,}225\ \text{mm}^2 + 1.0(0.4\ \text{kN/mm}^2)258\ \text{mm}^2$$
$$= 397\ \text{kN}$$

$$R_{nY} = 0.6F_yA_{gv} + U_{bs}F_uA_{nt}$$

$$R_{nY} = 0.6(36\ \text{k/in}^2)2.72\ \text{in}^2 + 1.0(58\ \text{k/in}^2)0.398\ \text{in}^2$$
$$= 81.8\ \text{k}$$

$$R_{nY} = 0.6(0.248\ \text{kN/mm}^2)1{,}748\ \text{mm}^2 + 1.0(0.4\ \text{kN/mm}^2)258\ \text{mm}^2$$
$$= 363\ \text{kN}$$

The second equation controls. Multiplying by $\phi = 0.75$, we get

$$\phi R_{nY} = 0.75R_{nY}$$
$$= 0.75(81.8\ \text{k})$$
$$= 61.4\ \text{k}$$

$$= 0.75(363\ \text{kN})$$
$$= 272\ \text{kN}$$

This is higher than T_u, which tells us we have adequate block shear strength.

Step 5: Compare Strength to Load

We see that each load transfer mechanism (failure mode) has greater capacity than demand. Our joint has adequate strength.

Step 6: Sketch Final Connection

Figure 8.26a shows our final connection geometry.

For interest sake, let's look at how much fillet it would take to carry the required force. From Table 8.12, we see we need to use at least a 3/16 in (5 mm) weld on a 3/8 in (9.5 mm) thick angle. And from Table 8.6, we see the capacity per inch for a 70 k/in^2 (483 MN/m^2) weld metal is

$$\phi r_{nW} = 4.18\ \text{k/in}$$

$$\phi r_{nW} = 0.768\ \text{kN/mm}$$

Dividing the total force by the weld strength per inch, we get the required weld length:

$$l_w = \frac{T_u}{\phi r_{nw}}$$

$$= \frac{48 \text{ k}}{4.18 \text{ k/in}} \qquad = \frac{214 \text{ kN}}{0.768 \text{ kN/mm}}$$

$$= 11.5 \text{ in} \qquad = 279 \text{ mm}$$

We'll round up to 12 in and 300 mm, and place half of this on each side of the angle, shown in Figure 8.26b.

And there you have it.

As a note, we have not yet designed the gusset plate. We need to check tension yield and rupture, along with bolt bearing. Multiplying l_G in Figure 8.26a by the gusset thickness, we get the gross area that is effective in carrying the tension force, and work from there.

8.7 WHERE WE GO FROM HERE

In this chapter, we have looked at the basic structural steel connections; bolts and welds, and their connecting elements. There are myriad iterations on how we use this information that allow us to design simple and complex connections. Looking at Parts 7 through 15 of the *Manual*, we see a great expansion on these ideas and their application to specific connection types.

Welding, like many topics in structural engineering, can consume a lifetime of study and practice. When forces acting on a weld become complex—say with tension, shear, moment, and torsion—it is helpful to treat the weld as a line. This method is documented in *Design of Welded Structures*,[8] by Omar Blodgett. Based on classical mechanics, this method computes weld section properties (A, S, J) as though the weld has no thickness. One then divides force, moment, and torsion by the appropriate section property, combines stresses, and backs out weld thickness. It is somewhat conservative compared to the inelastic methods now permitted in the code, but it is the only reasonable method for complex loadings on welds.

NOTES

1. SAE. Mechanical and Material Requirements for Externally Threaded Fasteners, J429 (Warrendale: SAE International, 2014).
2. AISC. Specification for Structural Joints using High-Strength Bolts (Chicago: American Institute of Steel Construction, 2009).
3. AWS. Structural Welding Code—Steel, D1.1 (Miami: American Welding Society, 2015); X AWS. Structural Welding Code—Sheet Metal, D1.3 (Miami: American Welding Society, 2008).
4. AWS. Structural Welding Code—Reinforcing Steel, D1.4 (Miami: American Welding Society, 2011).
5. AWS. Bridge Welding Code, D1.5 (Miami: American Welding Society, 2015).
6. AWS. Structural Welding Seismic, D1.8 (Miami: American Welding Society, 2016)
7. AISC, Specification for Structural Steel Buildings, AISC 360 (Chicago: American Institute of Steel Construction, 2016).
8. Blodgett, O.W. Design of Welded Structures, 12th Printing (Ohio: Lincoln Arc Welding Foundation, 1982).

Selected Section Properties

Appendix 1

Table A1.1. Wide flange shape section information

Section	A	d	t_w	b_f	t_f	I_x	Z_x	S_x	r_x	I_y	Z_y	r_y	Section
(Imperial)	(in^2)	(in)	(in)	(in)	(in)	(in^4)	(in^3)	(in^3)	(in)	(in^4)	(in^3)	(in)	(Metric)
W44×335	98.5	44.0	1.03	15.9	1.77	31,100	1,620	1,410	17.8	1,200	236	3.49	W1100×499
W40×593	174	43.0	1.79	16.7	3.23	50,400	2,760	2,340	17.0	2,520	481	3.80	W1000×883
W40×397	116	41.6	1.42	12.4	2.52	29,900	1,710	1,440	16.1	803	212	2.64	W1000×591
W40×149	43.8	38.2	0.630	11.8	0.830	9,800	598	513	15.0	229	62.2	2.29	W1000×222
W36×652	192	41.1	1.97	17.6	3.54	50,600	2,910	2,460	16.2	3,230	581	4.10	W920×970
W36×210	61.9	36.7	0.830	12.2	1.36	13,200	833	719	14.6	411	107	2.58	W920×313
W36×135	39.9	35.6	0.600	12.0	0.790	7,800	509	439	14.0	225	59.7	2.38	W920×201
W33×387	114	36.0	1.26	16.2	2.28	24,300	1,560	1,350	14.6	1,620	312	3.77	W840×576
W33×201	59.1	33.7	0.715	15.7	1.15	11,600	773	686	14.0	749	147	3.56	W840×299
W33×118	34.7	32.9	0.550	11.5	0.740	5,900	415	359	13.0	187	51.3	2.32	W840×176
W30×391	115	33.2	1.36	15.6	2.44	20,700	1,450	1,250	13.4	1,550	310	3.67	W760×582
W30×211	62.3	30.9	0.775	15.1	1.32	10,300	751	665	12.9	757	155	3.49	W760×314
W30×116	34.2	30.0	0.565	10.5	0.850	4,930	378	329	12.0	164	49.2	2.19	W760×173
W30×90	26.3	29.5	0.470	10.4	0.610	3,610	283	245	11.7	115	34.7	2.09	W760×134

Table A1.1 *continued*

Section	A	d	t_w	b_f	t_f	I_x	Z_x	S_x	r_x	I_y	Z_y	r_y	Section
(Imperial)	(in^2)	(in)	(in)	(in)	(in)	(in^4)	(in^3)	(in^3)	(in)	(in^4)	(in^3)	(in)	(Metric)
W27×539	159	32.5	1.97	15.3	3.54	25,600	1,890	1,570	12.7	2,110	437	3.65	W690×802
W27×217	63.9	28.4	0.830	14.1	1.50	8,910	711	627	11.8	704	154	3.32	W690×323
W27×84	24.7	26.7	0.460	10.0	0.640	2,850	244	213	10.7	106	33.2	2.07	W690×125
W24×370	109	28.0	1.52	13.7	2.72	13,400	1,130	957	11.1	1,160	267	3.27	W610×551
W24×162	47.8	25.0	0.705	13.0	1.22	5,170	468	414	10.4	443	105	3.05	W610×241
W24×103	30.3	24.5	0.550	9.00	0.980	3,000	280	245	10.0	119	41.5	1.99	W610×153
W24×55	16.2	23.6	0.395	7.01	0.505	1,350	134	114	9.11	29	13.3	1.34	W610×82
W21 × 201	59.3	23.0	0.910	12.6	1.63	5,310	530	461	9.47	542	133	3.02	W530×300
W21 × 122	35.9	21.7	0.600	12.4	0.960	2,960	307	273	9.09	305	75.6	2.92	W530×182
W21 × 73	21.5	21.2	0.455	8.30	0.740	1,600	172	151	8.64	70.6	26.6	1.81	W530×109
W21 × 50	14.7	20.8	0.380	6.53	0.535	984	110	94.5	8.18	24.9	12.2	1.30	W530×74
W18 × 311	91.6	22.3	1.52	12.0	2.74	6,970	754	624	8.72	795	207	2.95	W460×464
W18 × 158	46.3	19.7	0.810	11.3	1.44	3,060	356	310	8.12	347	94.8	2.74	W460×235
W18 × 71	20.9	18.5	0.495	7.64	0.810	1,170	146	127	7.50	60.3	24.7	1.70	W460×106
W18 × 40	11.8	17.9	0.315	6.02	0.525	612	78.4	68.4	7.21	19.1	10.0	1.27	W460×60
W18×35	10.3	17.7	0.300	6.00	0.425	510	66.5	57.6	7.04	15.3	8.06	1.22	W460×52

Table A1.1 *continued*

W16×100	29.4	17.0	0.585	10.4	0.985	1,490	198	175	7.10	186	54.9	2.51	W410×149
W16×57	16.8	16.4	0.430	7.12	0.715	758	105	92.2	6.72	43.1	18.9	1.60	W410×85
W16×31	9.13	15.9	0.275	5.53	0.440	375	54.0	47.2	6.41	12.4	7.0	1.17	W410×46.1
W16×26	7.68	15.7	0.250	5.50	0.345	301	44.2	38.4	6.26	9.59	5.48	1.12	W410×38.8
W14×730	215	22.4	3.07	17.9	4.91	14,300	1660	1280	8.17	4,720	816	4.69	W360×1086
W14×257	75.6	16.4	1.18	16.0	1.89	3,400	487	415	6.71	1,290	246	4.13	W360×382
W14×90	26.5	14.0	0.440	14.5	0.710	999	157	143	6.14	362	75.6	3.70	W360×134
W14×43	12.6	13.7	0.305	8.00	0.530	428	69.6	62.6	5.82	45.2	17.3	1.89	W360×64
W14×22	6.49	13.7	0.230	5.00	0.335	199	33.2	29.0	5.54	7.0	4.39	1.04	W360×32.9
W12×336	98.9	16.8	1.78	13.4	2.96	4,060	603	483	6.41	1,190	274	3.47	W310×500
W12×230	67.7	15.1	1.29	12.9	2.07	2,420	386	321	5.97	742	177	3.31	W310×342
W12×152	44.7	13.7	0.870	12.5	1.40	1,430	243	209	5.66	454	111	3.19	W310×226
W12×96	28.2	12.7	0.550	12.2	0.900	833	147	131	5.44	270	67.5	3.09	W310×143
W12×65	19.1	12.1	0.390	12.0	0.605	533	96.8	87.9	5.28	174	44.1	3.02	W310×97
W12×58	17.0	12.2	0.360	10.0	0.640	475	86.4	78.0	5.28	107	32.5	2.51	W310×86
W12×53	15.6	12.1	0.345	10.0	0.575	425	77.9	70.6	5.23	95.8	29.1	2.48	W310×79
W12×50	14.6	12.2	0.370	8.08	0.640	391	71.9	64.2	5.18	56.3	21.3	1.96	W310×74
W12×40	11.7	11.9	0.295	8.01	0.515	307	57.0	51.5	5.13	44.1	16.8	1.94	W310×60

Table A1.1 *continued*

Section	A	d	t_w	b_f	t_f	I_x	Z_x	S_x	r_x	I_y	Z_y	r_y	Section
(Imperial)	(in^2)	(in)	(in)	(in)	(in)	(in^4)	(in^3)	(in^3)	(in)	(in^4)	(in^3)	(in)	(Metric)
W12×35	10.3	12.5	0.300	6.56	0.520	285	51.2	45.6	5.25	24.5	11.5	1.54	W310×52
W12×26	7.65	12.2	0.230	6.49	0.380	204	37.2	33.4	5.17	17.3	8.17	1.51	W310×38.7
W12×19	5.57	12.2	0.235	4.01	0.350	130	24.7	21.3	4.82	3.76	2.98	0.822	W310×28.3
W12×14	4.16	11.9	0.200	3.97	0.225	88.6	17.4	14.9	4.62	2.36	1.90	0.753	W310×21
W10×112	32.9	11.4	0.755	10.4	1.25	716	147	126	4.66	236	69.2	2.68	W250×167
W10×49	14.4	10.0	0.340	10.0	0.560	272	60.4	54.6	4.35	93.4	28.3	2.54	W250×73
W10×33	9.71	9.73	0.290	7.96	0.435	171	38.8	35.0	4.19	36.6	14.0	1.94	W250×49.1
W10×22	6.49	10.2	0.240	5.75	0.360	118	26.0	23.2	4.27	11.4	6.10	1.33	W250×32.7
W10×12	3.54	9.87	0.190	3.96	0.210	53.8	12.6	10.9	3.90	2.18	1.74	0.785	W250×17.9
W8×67	19.7	9.00	0.570	8.28	0.935	272	70.1	60.4	3.72	88.6	32.7	2.12	W200×100
W8×24	7.08	7.93	0.245	6.50	0.400	82.7	23.1	20.9	3.42	18.3	8.57	1.61	W200×35.9
W8×18	5.26	8.14	0.230	5.25	0.330	61.9	17.0	15.2	3.43	7.97	4.66	1.23	W200×26.6
W8×10	2.96	7.89	0.170	3.94	0.205	30.8	8.87	7.81	3.22	2.09	1.66	0.841	W200×15
W6×25	7.34	6.38	0.320	6.08	0.455	53.4	18.9	16.7	2.70	17.1	8.56	1.52	W150×37.1
W5×19	5.56	5.15	0.270	5.03	0.430	26.3	11.6	10.2	2.17	9.13	5.53	1.28	W130×28.1
W4×13	3.83	4.16	0.280	4.06	0.345	11.3	6.28	5.46	1.72	3.86	2.92	1.00	W100×19.3

Source: AISC Steel Construction Manual, 14th edition

1) Not all W sections are shown.

2) M, S, and HP sections not shown. See Steel manual.

Table A1.1m Wide flange shape section information

Section	A	d	t_w	b_f	t_f	I_x / 10^6	Z_x / 10^3	S_x / 10^3	r_x	I_y / 10^6	Z_y / 10^3	r_y	Section
(Metric)	(mm^2)	(mm)	(mm)	(mm)	(mm)	(mm^4)	(mm^3)	(mm^3)	(mm)	(mm^4)	(mm^3)	(mm)	(Imperial)
W1100x499	63,500	1,120	26.2	404	45.0	12,900	26,500	23,100	452	499	3,870	88.6	W44x335
W1000x883	112,000	1,090	45.5	424	82.0	21,000	45,200	38,300	432	1,050	7,880	96.5	W40x593
W1000x584	74,800	1,060	36.1	315	64.0	12,400	28,000	23,600	409	334	3,470	67.1	W40x392
W1000x222	28,300	970	16.0	300	21.1	4,080	9,800	8,410	381	95.3	1,020	58.2	W40x149
W920x970	124,000	1,040	50.0	447	89.9	21,100	47,700	40,300	411	1,340	9,520	104	W36x652
W920x313	39,900	932	21.1	310	34.5	5,490	13,700	11,800	371	171	1,750	65.5	W36x210
W920x201	25,700	904	15.2	305	20.1	3,250	8,340	7,190	356	93.7	978	60.5	W36x135
W840x576	73,500	914	32.0	411	57.9	10,100	25,600	22,100	371	674	5,110	95.8	W33x387
W840x299	38,100	856	18.2	399	29.2	4,830	12,700	11,200	356	312	2,410	90.4	W33x201
W840x176	22,400	836	14.0	292	18.8	2,460	6,800	5,880	330	77.8	841	58.9	W33x118
W760x582	74,200	843	34.5	396	62.0	8,620	23,800	20,500	340	645	5,080	93.2	W30x391
W760x314	40,200	785	19.7	384	33.5	4,290	12,300	10,900	328	315	2,540	88.6	W30x211
W760x173	22,100	762	14.4	267	21.6	2,050	6,190	5,390	305	68	806	55.6	W30x116
W760x134	17,000	749	11.9	264	15.5	1,500	4,640	4,010	297	47.9	569	53.1	W30x90

Table A1.1m ***continued***

Section	A	d	t_w	b_f	t_f	I_x / 10^6	Z_x / 10^3	S_x / 10^3	r_x	I_y / 10^6	Z_y / 10^3	r_y	Section
(Metric)	(mm^2)	(mm)	(mm)	(mm)	(mm)	(mm^4)	(mm^3)	(mm^3)	(mm)	(mm^4)	(mm^3)	(mm)	*(Imperial)*
W690x802	103,000	826	50.0	389	89.9	10,700	31,000	25,700	323	878	7,160	92.7	W27x539
W690x323	41,200	721	21.1	358	38.1	3,710	11,700	10,300	300	293	2,520	84.3	W27x217
W690x125	15,900	678	11.7	254	16.3	1,190	4,000	3,490	272	44.1	544	52.6	W27x84
W610x551	70,300	711	38.6	348	69.1	5,580	18,500	15,700	282	483	4,380	83.1	W24x370
W610x241	30,800	635	17.9	330	31.0	2,150	7,670	6,780	264	184	1,720	77.5	W24x162
W610x153	19,500	622	14.0	229	24.9	1,250	4,590	4,010	254	49.5	680	50.5	W24x103
W610x82	10,500	599	10.0	178	12.8	562	2,200	1,870	231	12.1	218	34.0	W24x55
W530x300	38,300	584	23.1	320	41.4	2,210	8,690	7,550	241	226	2,180	76.7	W21x201
W530x182	23,200	551	15.2	315	24.4	1,230	5,030	4,470	231	127	1,240	74.2	W21x122
W530x109	13,900	538	11.6	211	18.8	666	2,820	2,470	219	29.4	436	46.0	W21x73
W530x74	9,480	528	9.65	166	13.6	410	1,800	1,550	208	10.4	200	33.0	W21x50
W460x464	59,100	566	38.6	305	69.6	2,900	12,400	10,200	221	331	3,390	74.9	W18x311
W460x235	29,900	500	20.6	287	36.6	1,270	5,830	5,080	206	144	1,550	69.6	W18x158
W460x106	13,500	470	12.6	194	20.6	487	2,390	2,080	191	25.1	405	43.2	W18x71
W460x60	7,610	455	8.00	153	13.3	255	1,280	1,120	183	8.0	164	32.3	W18x40
W460x52	6,650	450	7.62	152	10.8	212	1,090	944	179	6.37	132	31.0	W18x35

Table A1.m *continued*

W410x149	19,000	432	14.9	264	25.0	620	3,240	2,870	180	77.4	900	63.8	W16x100
W410x85	10,800	417	10.9	181	18.2	316	1,720	1,510	171	17.9	310	40.6	W16x57
W410x46.1	5,890	404	6.99	140	11.2	156	885	773	163	5.2	115	29.7	W16x31
W410x38.8	4,950	399	6.35	140	8.76	125	724	629	159	3.99	89.8	28.4	W16x26
W360x1086	139,000	569	78.0	455	125	5,950	27,200	21,000	208	1,960	13,400	119	W14x730
W360x382	48,800	417	30.0	406	48.0	1,420	7,980	6,800	170	537	4,030	105	W14x257
W360x134	17,100	356	11.2	368	18.0	416	2,570	2,340	156	151	1,240	94.0	W14x90
W360x64	8,130	348	7.75	203	13.5	178	1,140	1,030	148	18.8	283	48.0	W14x43
W360x32.9	4,190	348	5.84	127	8.51	82.8	544	475	141	2.91	71.9	26.4	W14x22
W310x500	63,800	427	45.2	340	75.2	1,690	9,880	7,910	163	495	4,490	88.1	W12x336
W310x342	43,700	384	32.8	328	52.6	1,010	6,330	5,260	152	309	2,900	84.1	W12x230
W310x226	28,800	348	22.1	318	35.6	595	3,980	3,420	144	189	1,820	81.0	W12x152
W310x143	18,200	323	14.0	310	22.9	347	2,410	2,150	138	112	1,110	78.5	W12x96
W310x97	12,300	307	9.91	305	15.4	222	1,590	1,440	134	72.4	723	76.7	W12x65
W310x86	11,000	310	9.14	254	16.3	198	1,420	1,280	134	44.5	533	63.8	W12x58
W310x79	10,100	307	8.76	254	14.6	177	1,280	1,160	133	39.9	477	63.0	W12x53
W310x74	9,420	310	9.40	205	16.3	163	1,180	1,050	132	23.4	349	49.8	W12x50
W310x60	7,550	302	7.49	203	13.1	128	934	844	130	18.4	275	49.3	W12x40

Table A1.1m *continued*

Section	A	d	t_w	b_f	t_f	I_x / 10^6	Z_x / 10^3	S_x / 10^3	r_x	I_y / 10^6	Z_y / 10^3	r_y	*Section*
(Metric)	(mm^2)	(mm)	(mm)	(mm)	(mm)	(mm^4)	(mm^3)	(mm^3)	(mm)	(mm^4)	(mm^3)	(mm)	*(Imperial)*
W310x52	6,650	318	7.62	167	13.2	119	839	747	133	10.2	188	39.1	W12x35
W310x38.7	4,940	310	5.84	165	9.65	84.9	610	547	131	7.20	134	38.4	W12x26
W310x28.3	3,590	310	5.97	102	8.89	54.1	405	349	122	1.57	48.8	20.9	W12x19
W310x21	2,680	302	5.08	101	5.72	36.9	285	244	117	0.982	31.1	19.1	W12x14
W250x167	21,200	290	19.2	264	31.8	298	2,410	2,060	118	98.2	1130	68.1	W10x112
W250x73	9,290	254	8.64	254	14.2	113	990	895	110	38.9	464	64.5	W10x49
W250x49.1	6,260	247	7.37	202	11.0	71.2	636	574	106	15.2	229	49.3	W10x33
W250x32.7	4,190	259	6.10	146	9.14	49.1	426	380	108	4.75	100	33.8	W10x22
W250x17.9	2,280	251	4.83	101	5.33	22.4	206	179	99.1	0.907	28.5	19.9	W10x12
W200x100	12,700	229	14.5	210	23.7	113	1,150	990	94.5	36.9	536	53.8	W8x67
W200x35.9	4,570	201	6.22	165	10.2	34.4	379	342	86.9	7.62	140	40.9	W8x24
W200x26.6	3,390	207	5.84	133	8.38	25.8	279	249	87.1	3.32	76.4	31.2	W8x18
W200x15	1,910	200	4.32	100	5.21	12.8	145	128	81.8	0.870	27.2	21.4	W8x10
W150x37.1	4,740	162	8.13	154	11.6	22.2	310	274	68.6	7.12	140	38.6	W6x25
W130x28.1	3,590	131	6.86	128	10.9	10.9	190	167	55.1	3.80	90.6	32.5	W5x19
W100x19.3	2,470	106	7.11	103	8.76	4.70	103	89.5	43.7	1.61	47.9	25.4	W4x13

Source: AISC Steel Construction Manual, 14th edition

1) Not all W sections are shown. 2) M, S, and HP sections not shown. See Steel manual.
3) I/10^6, multiply table value by 10^6 4. Z or S/103, multiply table value by 103

Table A1.2. Channel shape section information

Section	W	A	d	t_w	b_f	t_f	I_x	Z_x	S_x	r_x	I_y	Z_y	r_y	Section
(Imperial)	*(lb/ft)*	*(in^2)*	*(in)*	*(in)*	*(in)*	*(in)*	*(in^4)*	*(in^3)*	*(in^3)*	*(in)*	*(in^4)*	*(in^3)*	*(in)*	*(Metric)*
C15x50	50.0	14.7	15.0	0.716	3.72	0.650	404	68.5	53.8	5.24	11.0	8.14	0.865	C380x74
C15x33.9	33.9	10.0	15.0	0.400	3.40	0.650	315	50.8	42.0	5.61	8.07	6.19	0.901	C380x50.4
C12x30	30.0	8.81	12.0	0.510	3.17	0.501	162	33.8	27.0	4.29	5.12	4.32	0.762	C310x45
C12x20.7	20.7	6.08	12.0	0.282	2.94	0.501	129	25.6	21.5	4.61	3.86	3.47	0.797	C310x30.8
C10x30	30.0	8.81	10.0	0.673	3.03	0.436	103	26.7	20.7	3.43	3.93	3.78	0.668	C250x45
C10x15.3	15.3	4.48	10.0	0.240	2.60	0.436	67.3	15.9	13.5	3.88	2.27	2.34	0.711	C250x22.8
C9x20	20.0	5.87	9.00	0.448	2.65	0.413	60.9	16.9	13.5	3.22	2.41	2.46	0.640	C230x30
C9x13.4	13.4	3.94	9.00	0.233	2.43	0.413	47.8	12.6	10.6	3.48	1.75	1.94	0.666	C230x19.9
C8x18.75	18.75	5.51	8.00	0.487	2.53	0.390	43.9	13.9	11.0	2.82	1.97	2.17	0.598	C200x27.9
C8x11.5	11.5	3.37	8.00	0.220	2.26	0.390	32.5	9.63	8.14	3.11	1.31	1.57	0.623	C200x17.1
C7x9.8	9.80	2.87	7.00	0.210	2.09	0.366	21.2	7.19	6.07	2.72	0.957	1.26	0.578	C180x14.6
C6x8.2	8.20	2.39	6.00	0.200	1.92	0.343	13.1	5.16	4.35	2.34	0.687	0.987	0.536	C150x12.2
C5x6.7	6.70	1.97	5.00	0.190	1.75	0.320	7.48	3.55	2.99	1.95	0.470	0.757	0.489	C130x10.4
C4x4.5	4.50	1.38	4.00	0.125	1.58	0.296	3.65	2.12	1.83	1.63	0.289	0.531	0.457	C100x6.7
C3x3.5	3.50	1.09	3.00	0.132	1.37	0.273	1.57	1.24	1.04	1.20	0.169	0.364	0.394	C75x5.2

Source: AISC Steel Construction Manual, 14th edition

1) Not all C sections are shown. 2) Miscellaneous Channel (MC) sections not shown. See Steel manual.

Table A1.2. Channel sshape section information

Section	W	A	d	t_w	b_f	t_f	I_x / 10^6	Z_x / 10^3	S_x / 10^3	r_x	I_y / 10^6	Z_y / 10^3	r_y	Section
(Metric)	(kg/m)	(mm^2)	(mm)	(mm)	(mm)	(mm)	(mm^4)	(mm^3)	(mm^3)	(mm)	(mm^4)	(mm^3)	(mm)	(Imperial)
C380x74	74.0	9,480	381	18.2	94.5	16.5	168	1120	882	133	4.58	133	22.0	C15x50
C380x50.4	50.4	6,450	381	10.2	86.4	16.5	131	832	688	142	3.36	101	22.9	C15x33.9
C310x45	45.0	5,680	305	13.0	80.5	12.7	67.4	554	442	109	2.13	70.8	19.4	C12x30
C310x30.8	30.8	3,920	305	7.16	74.7	12.7	53.7	420	352	117	1.61	56.9	20.2	C12x20.7
C250x45	45.0	5,680	254	17.1	77.0	11.1	42.9	438	339	87.1	1.64	61.9	17.0	C10x30
C250x22.8	22.8	2,890	254	6.10	66.0	11.1	28.0	261	221	98.6	0.945	38.3	18.1	C10x15.3
C230x30	30.0	3,790	229	11.4	67.3	10.5	25.3	277	221	81.8	1.00	40.3	16.3	C9x20
C230x19.9	19.9	2,540	229	5.92	61.7	10.5	19.9	206	174	88.4	0.728	31.8	16.9	C9x13.4
C200x27.9	27.9	3,550	203	12.4	64.3	9.91	18.3	228	180	71.6	0.820	35.6	15.2	C8x18.75
C200x17.1	17.1	2,170	203	5.59	57.4	9.91	13.5	158	133	79.0	0.545	25.7	15.8	C8x11.5
C180x14.6	14.6	1,850	178	5.33	53.1	9.30	8.82	118	99.5	69.1	0.398	20.6	14.7	C7x9.8
C150x12.2	12.2	1,540	152	5.08	48.8	8.71	5.45	84.6	71.3	59.4	0.286	16.2	13.6	C6x8.2
C130x10.4	10.4	1,270	127	4.83	44.5	8.13	3.11	58.2	49.0	49.5	0.196	12.4	12.4	C5x6.7
C100x6.7	6.70	890	102	3.18	40.1	7.52	1.52	34.7	30.0	41.4	0.120	8.70	11.6	C4x4.5
C75x5.2	5.20	703	76.2	3.35	34.8	6.93	0.653	20.3	17.0	30.5	0.0703	5.96	10.0	C3x3.5

Source: AISC Steel Construction Manual, 14th edition.

1) Not all C sections are shown. 2) Miscellaneous Channel (MC) sections not shown. See Steel manual. 3) $I/10^6$, multiply table value by 10^6 4. Z or $S/10^3$, multiply table value by 10^3

Table A1.3 Angle shape section information

Section (Imperial)	W (lb/ft)	A (in^2)	d, b (in)	t (in)	I (in^4)	Z (in^3)	S (in^3)	r (in)	I_z (in^4)	S_z (in^3)	r_z (in)	Section (Metric)
L8x8x7/8	45.0	13.3	8.00	0.875	79.7	25.3	14.0	2.45	32.7	10.0	1.57	L203x203x22.2
L8x8x1/2	26.4	7.84	8.00	0.500	48.8	15.1	8.36	2.49	19.8	6.44	1.59	L203x203x12.7
L6x6x1	37.4	11.0	6.00	1.00	35.4	15.4	8.55	1.79	14.9	5.70	1.17	L152x152x25.4
L6x6x9/16	21.9	6.45	6.00	0.563	22.0	9.18	5.12	1.85	8.90	3.73	1.18	L152x152x14.3
L6x6x5/16	12.4	3.67	6.00	0.313	13.0	5.26	2.95	1.88	5.20	2.30	1.19	L152x152x7.9
L5x5x7/8	27.2	8.00	5.00	0.875	17.8	9.31	5.16	1.49	7.60	3.43	0.971	L127x127x22.2
L5x5x7/16	14.3	4.22	5.00	0.438	10.0	5.00	2.78	1.54	4.04	2.06	0.983	L127x127x11.1
L5x5x5/16	10.3	3.07	5.00	0.313	7.44	3.65	2.04	1.56	3.00	1.58	0.990	L127x127x7.9
L4x4x3/4	18.5	5.44	4.00	0.750	7.62	5.02	2.79	1.18	3.25	1.81	0.774	L102x102x19
L4x4x1/2	12.8	3.75	4.00	0.500	5.52	3.50	1.96	1.21	2.25	1.35	0.776	L102x102x12.7
L4x4x3/8	9.80	2.86	4.00	0.375	4.32	2.69	1.50	1.23	1.73	1.08	0.779	L102x102x9.5
L4x4x1/4	6.60	1.93	4.00	0.250	3.00	1.82	1.03	1.25	1.19	0.776	0.783	L102x102x6.4

Table A1.3 ***continued***

Section	W	A	d, b	t	I	Z	S	r	I_z	S_z	r_z	*Section*
(Imperial)	*(lb/ft)*	*(in^2)*	*(in)*	*(in)*	*(in^4)*	*(in^3)*	*(in^3)*	*(in)*	*(in^4)*	*(in^3)*	*(in)*	*(Metric)*
L3x3x1/2	9.40	2.76	3.00	0.500	2.20	1.91	1.06	0.895	0.922	0.703	0.580	L76x76x12.7
L3x3x3/8	7.20	2.11	3.00	0.375	1.75	1.48	0.825	0.910	0.716	0.570	0.581	L76x76x9.5
L3x3x1/4	4.90	1.44	3.00	0.250	1.23	1.02	0.569	0.926	0.490	0.415	0.585	L76x76x6.4
L3x3x3/16	3.71	1.09	3.00	0.188	0.948	0.774	0.433	0.933	0.373	0.326	0.586	L76x76x4.8
L2x2x1/4	3.19	0.944	2.00	0.250	0.346	0.440	0.244	0.605	0.142	0.171	0.387	L51x51x6.4
L2x2x1/8	1.65	0.491	2.00	0.125	0.189	0.230	0.129	0.620	0.0756	0.0994	0.391	L51x51x3.2

Source: AISC Steel Construction Manual, 14th edition

1) Not all L sections are shown.
2) Unequal angle and double angle sections not shown. See Steel manual.

Table A1.3m Angle shape section information

Section	W	A	d, b	t	$I / 10^6$	$Z / 10^3$	$S / 10^3$	r	$I_z / 10^6$	$S_z / 10^3$	r_z	Section
(Metric)	(kg/m)	(mm^2)	(mm)	(mm)	(mm^4)	(mm^3)	(mm^3)	(mm)	(mm^4)	(mm^3)	(mm)	(Imperial)
L203x203x22.2	67.0	8,580	203	22.2	33.2	415	229	62.2	13.6	164	39.9	L8x8x7/8
L203x203x12.7	39.3	5,060	203	12.7	20.3	247	137	63.2	8.24	106	40.4	L8x8x1/2
L152x152x25.4	55.7	7,100	152	25.4	14.7	252	140	45.5	6.20	93.4	29.7	L6x6x1
L152x152x14.3	32.6	4,160	152	14.3	9.16	150	83.9	47.0	3.70	61.1	30.0	L6x6x9/16
L152x152x7.9	18.5	2,370	152	7.94	5.41	86.2	48.3	47.8	2.16	37.7	30.2	L6x6x5/16
L127x127x22.2	40.5	5,160	127	22.2	7.41	153	84.6	37.8	3.16	56.2	24.7	L5x5x7/8
L127x127x11.1	21.3	2,720	127	11.1	4.16	81.9	45.6	39.1	1.68	33.8	25.0	L5x5x7/16
L127x127x7.9	15.3	1,980	127	7.94	3.10	59.8	33.4	39.6	1.25	25.9	25.1	L5x5x5/16
102x102x19	27.5	3,510	102	19.1	3.17	82.3	45.7	30.0	1.35	29.7	19.7	L4x4x3/4
L102x102x12.7	19.0	2,420	102	12.7	2.30	57.4	32.1	30.7	0.937	22.1	19.7	L4x4x1/2
L102x102x9.5	14.6	1,850	102	9.53	1.80	44.1	24.6	31.2	0.720	17.7	19.8	L4x4x3/8
L102x102x6.4	9.80	1,250	102	6.35	1.25	29.8	16.9	31.8	0.495	12.7	19.9	L4x4x1/4

Table A1.3 ***continued***

Section	*W*	*A*	*d, b*	*t*	$I / 10^6$	$Z / 10^3$	$S / 10^3$	r	$I_z / 10^6$	$S_z / 10^3$	r_z	*Section*
(Metric)	*(kg/m)*	*(mm²)*	*(mm)*	*(mm)*	*(mm⁴)*	*(mm³)*	*(mm³)*	*(mm)*	*(mm⁴)*	*(mm³)*	*(mm)*	*(Imperial)*
L76x76x12.7	14.0	1,780	76.2	12.7	0.916	31.3	17.4	22.7	0.384	11.5	14.7	L3x3x1/2
L76x76x9.5	10.7	1,360	76.2	9.53	0.728	24.3	13.5	23.1	0.298	9.34	14.8	L3x3x3/8
L76x76x6.4	7.30	929	76.2	6.35	0.512	16.7	9.32	23.5	0.204	6.80	14.9	L3x3x1/4
L76x76x4.8	5.50	703	76.2	4.76	0.395	12.7	7.10	23.7	0.155	5.34	14.9	L3x3x3/16
L51x51x6.4	4.70	609	50.8	6.35	0.144	7.21	4.00	15.4	0.0591	2.80	9.83	L2x2x1/4
L51x51x3.2	2.40	317	50.8	3.18	0.0787	3.77	2.11	15.7	0.0315	1.63	9.93	L2x2x1/8

Source: AISC Steel Construction Manual, 14th edition

1) Not all L sections are shown.
2) Unequal angle and double angle sections not shown. See Steel manual.
3) $I/10^6$, multiply table value by 10^6.
4) Z or $S/10^3$, multiply table value by 10^3.

Table A1.4 Hollow structural section shape section information

Section	*W*	*A*	*d, b*	*t*	*I*	*Z*	*S*	*r*	*C*	*Section*
(Imperial)	*(lb/ft)*	*(in^2)*	*(in)*	*(in)*	*(in^4)*	*(in^3)*	*(in^3)*	*(in)*	*(in^3)*	*(Metric)*
HSS16x16x5/8	127.37	35.0	16.0	0.581	1370	200	171	6.25	276	HSS406.4x406.4x15.9
HSS16x16x3/8	78.52	21.5	16.0	0.349	873	126	109	6.37	171	HSS406.4x406.4x9.5
HSS14x14x5/8	110.36	30.3	14.0	0.581	897	151	128	5.44	208	HSS355.6x355.6x15.9
HSS14x14x3/8	68.31	18.7	14.0	0.349	577	95.4	82.5	5.55	130	HSS355.6x355.6x9.5
HSS12x12x5/8	93.34	25.7	12.0	0.581	548	109	91.4	4.62	151	HSS304.8x304.8x15.9
HSS12x12x3/8	58.1	16.0	12.0	0.349	357	69.2	59.5	4.73	94.6	HSS304.8x304.8x9.5
HSS10x10x5/8	76.33	21.0	10.0	0.581	304	73.2	60.8	3.80	102	HSS254x254x15.9
HSS10x10x3/8	47.9	13.2	10.0	0.349	202	47.2	40.4	3.92	64.8	HSS254x254x9.5
HSS8x8x5/8	59.32	16.4	8.00	0.581	146	44.7	36.5	2.99	63.2	HSS203.2x203.2x15.9
HSS8x8x3/8	37.69	10.4	8.00	0.349	100	29.4	24.9	3.10	40.7	HSS203.2x203.2x9.5
HSS8x8x1/4	25.82	7.10	8.00	0.250	70.7	20.5	17.7	3.15	28.1	HSS203.2x203.2x6.4
HSS7x7x5/8	50.81	14.0	7.00	0.581	93.4	33.1	26.7	2.58	47.1	HSS177.8x177.8x15.9
HSS7x7x3/8	32.58	8.97	7.00	0.349	65.0	22.1	18.6	2.69	30.7	HSS177.8x177.8x9.5
HSS7x7x1/4	22.42	6.17	7.00	0.233	46.5	15.5	13.3	2.75	21.3	HSS177.8x177.8x6.4

Table A1.4 ***continued***

Section	*W*	*A*	*d, b*	*t*	*I*	*Z*	*S*	*r*	*C*	*Section*
(Imperial)	*(lb/ft)*	*(in^2)*	*(in)*	*(in)*	*(in^4)*	*(in^3)*	*(in^3)*	*(in)*	*(in^3)*	*(Metric)*
HSS6x6x5/8	42.3	11.7	6.00	0.581	55.2	23.2	18.4	2.17	33.4	HSS152.4x152.4x15.9
HSS6x6x3/8	27.48	7.58	6.00	0.349	39.5	15.8	13.2	2.28	22.1	HSS152.4x152.4x9.5
HSS6x6x1/4	19.02	5.24	6.00	0.233	28.6	11.2	9.54	2.34	15.4	HSS152.4x152.4x6.4
HSS5x5x3/8	22.37	6.18	5.00	0.349	21.7	10.6	8.68	1.87	14.9	HSS127x127x9.5
HSS5x5x1/4	15.62	4.30	5.00	0.233	16.0	7.61	6.41	1.93	10.5	HSS127x127x6.4
HSS4x4x3/8	17.27	4.78	4.00	0.349	10.3	6.39	5.13	1.47	9.14	HSS101.6x101.6x9.5
HSS4x4x1/4	12.21	3.37	4.00	0.233	7.80	4.69	3.90	1.52	6.56	HSS101.6x101.6x6.4
HSS3x3x1/4	8.81	2.44	3.00	0.233	3.02	2.48	2.01	1.11	3.52	HSS76.2x76.2x6.4
HSS3x3x1/8	4.75	1.30	3.00	0.116	1.78	1.40	1.19	1.17	1.92	HSS76.2x76.2x3.2

Source: AISC Steel Construction Manual, 14th edition

1) Not all L sections are shown.
2) Rectangular and round sections not shown. See Steel manual.

Table A1.4m Hollow structural section shape section information

Section (Metric)	W (kg/m)	A (mm^2)	d, b (mm)	t (mm)	I / 10^6 (mm^4)	Z / 10^3 (mm^3)	S / 10^3 (mm^3)	r (mm)	C / 10^3 (mm^3)	Section (Imperial)
HSS406.4x406.4x15.9	189	22,600	406	14.8	570	3,280	2,800	159	4,523	HSS16x16x5/8
HSS406.4x406.4x9.5	117	13,900	406	8.86	363	2,060	1,790	162	2,802	HSS16x16x3/8
HSS355.6x355.6x15.9	164	19,500	356	14.8	373	2,470	2,100	138	3,409	HSS14x14x5/8
HSS355.6x355.6x9.5	102	12,100	356	8.86	240	1,560	1,350	141	2,130	HSS14x14x3/8
HSS304.8x304.8x15.9	139	16,600	305	14.8	228	1,790	1,500	117	2,474	HSS12x12x5/8
HSS304.8x304.8x9.5	86.3	10,300	305	8.86	149	1,130	975	120	1,550	HSS12x12x3/8
HSS254x254x15.9	113	13,500	254	14.8	127	1,200	996	96.5	1,671	HSS10x10x5/8
HSS254x254x9.5	71.2	8,520	254	8.86	84.1	773	662	99.6	1,062	HSS10x10x3/8
HSS203.2x203.2x15.9	88	10,600	203	14.8	60.8	733	598	75.9	1,040	HSS8x8x5/8
HSS203.2x203.2x9.5	56.0	6,710	203	8.86	41.6	482	408	78.7	667	HSS8x8x3/8
HSS203.2x203.2x6.4	38.4	4,580	203	5.92	29.4	336	290	80.0	460	HSS8x8x1/4
HSS177.8x177.8x15.9	75.3	9,030	178	14.8	38.9	542	438	65.5	772	HSS7x7x5/8
HSS177.8x177.8x9.5	48.4	5,790	178	8.86	27.1	362	305	68.3	503	HSS7x7x3/8
HSS177.8x177.8x6.4	33.3	3,980	178	5.92	19.4	254	218	69.9	349	HSS7x7x1/4

Table A1.4m ***continued***

Section	*W*	*A*	*d, b*	*t*	*I / 10⁶*	*Z / 10³*	*S / 10³*	r	*C / 10³*	*Section*
(Metric)	*(kg/m)*	*(mm²)*	*(mm)*	*(mm)*	*(mm⁴)*	*(mm³)*	*(mm³)*	*(mm)*	*(mm³)*	*(Imperial)*
HSS152.4x152.4x15.9	62.6	7,550	152	14.8	23.0	380	302	55.1	547	HSS6x6x5/8
HSS152.4x152.4x9.5	40.8	4,890	152	8.86	16.4	259	216	57.9	362	HSS6x6x3/8
HSS152.4x152.4x6.4	28.3	3,380	152	5.92	11.9	184	156	59.4	252	HSS6x6x1/4
HSS127x127x9.5	33.2	3,990	127	8.86	9.03	174	142	47.5	244	HSS5x5x3/8
HSS127x127x6.4	23.2	2,770	127	5.92	6.66	125	105	49.0	172	HSS5x5x1/4
HSS101.6x101.6x9.5	25.6	3,080	102	8.86	4.29	105	84.1	37.3	150	HSS4x4x3/8
HSS101.6x101.6x6.4	18.1	2,170	102	5.92	3.25	76.9	63.9	38.6	107	HSS4x4x1/4
HSS76.2x76.2x6.4	13.1	1,570	76.2	5.92	1.26	40.6	32.9	28.2	57.7	HSS3x3x1/4
HSS76.2x76.2x3.2	7.06	839	76.2	2.95	0.741	22.9	19.5	29.7	31.5	HSS3x3x1/8

Source: AISC Steel Construction Manual, 14th edition

1) Not all HSS sections are shown.
2) Rectangular and round sections not shown. See Steel manual.
3) $I/10^6$, multiply table value by 10^6
4) Z, S, or $C/10^3$, multiply table value by 10^3

Table A1.5 Pipe shape section information

Section	*W*	*A*	*OD*	*ID*	t_{des}	I	Z	S	r	*Section*
(Imperial)	*(lb/ft)*	*(in^2)*	*(in)*	*(in)*	*(in)*	*(in^4)*	*(in^3)*	*(in^3)*	*(in)*	*(Metric)*
Standard										
Pipe12- STD	49.6	13.7	12.8	12.0	0.349	262	53.7	41.0	4.39	Pipe310STD
Pipe10- STD	40.5	11.5	10.8	10.0	0.340	151	36.9	28.1	3.68	Pipe254STD
Pipe 8- STD	28.6	7.85	8.63	7.98	0.300	68.1	20.8	15.8	2.95	Pipe203STD
Pipe 6- STD	19.0	5.20	6.63	6.07	0.261	26.5	10.6	7.99	2.25	Pipe152STD
Pipe 5- STD	14.6	4.01	5.56	5.05	0.241	14.3	6.83	5.14	1.88	Pipe127STD
Pipe 4- STD	10.8	2.96	4.50	4.03	0.221	6.82	4.05	3.03	1.51	Pipe102STD
Pipe 3- STD	7.58	2.07	3.50	3.07	0.201	2.85	2.19	1.63	1.17	Pipe75STD
Pipe 2–1/2- STD	5.80	1.61	2.88	2.47	0.189	1.45	1.37	1.01	0.952	Pipe64STD
Pipe 2- STD	3.66	1.02	2.38	2.07	0.143	0.627	0.713	0.528	0.791	Pipe51STD
Pipe 1–1/2- STD	2.72	0.749	1.90	1.61	0.135	0.293	0.421	0.309	0.626	Pipe38STD
Pipe 1–1/4- STD	2.27	0.625	1.66	1.38	0.130	0.184	0.305	0.222	0.543	Pipe32STD

Table A1.5 *continued*

Section	*W*	*A*	*OD*	*ID*	t_{des}	I	Z	S	r	*Section*
(Imperial)	*(lb/ft)*	*(in^2)*	*(in)*	*(in)*	*(in)*	*(in^4)*	*(in^3)*	*(in^3)*	*(in)*	*(Metric)*
Extra Strong										
Pipe12-XStrong	65.5	17.5	12.8	11.8	0.465	339	70.2	53.2	4.35	Pipe310XS
Pipe10-XStrong	54.8	15.1	10.8	9.75	0.465	199	49.2	37.0	3.64	Pipe254XS
Pipe8-XStrong	43.4	11.9	8.63	7.63	0.465	100	31.0	23.1	2.89	Pipe203XS
Pipe6-XStrong	28.6	7.83	6.63	5.76	0.403	38.3	15.6	11.6	2.20	Pipe152XS
Pipe5-XStrong	20.8	5.73	5.56	4.81	0.349	19.5	9.50	7.02	1.85	Pipe127XS
Pipe4-XStrong	15.0	4.14	4.50	3.83	0.315	9.12	5.53	4.05	1.48	Pipe102XS
Pipe3-XStrong	10.3	2.83	3.50	2.90	0.280	3.70	2.91	2.11	1.14	Pipe75XS

Source: AISC Steel Construction Manual, 14th edition

1) Double extra strong sections not shown. See Steel manual.

Table A1.5m Pipe shape section information

Section	*W*	*A*	*OD*	*ID*	t_{des}	*I / 10⁶*	*Z / 10³*	*S / 10³*	r	*Section*
(Metric)	*(kg/m)*	(mm^2)	*(mm)*	*(mm)*	*(mm)*	(mm^4)	(mm^3)	(mm^3)	*(mm)*	*(Imperial)*
Standard										
Pipe310- STD	73.8	8,840	325	305	8.86	109	880	672	112	Pipe12- STD
Pipe254- STD	60.2	7,420	274	254	8.64	62.9	605	460	93.5	Pipe10- STD
Pipe203- STD	42.5	5,060	219	203	7.62	28.3	341	259	74.9	Pipe 8- STD
Pipe152- STD	28.3	3,350	168	154	6.63	11.0	174	131	57.2	Pipe 6- STD
Pipe127- STD	21.7	2,590	141	128	6.12	5.95	112	84.2	47.8	Pipe 5- STD
Pipe102- STD	16.1	1,910	114	102	5.61	2.84	66.4	49.7	38.4	Pipe 4- STD
Pipe75- STD	11.3	1,340	88.9	78.0	5.11	1.19	35.9	26.7	29.7	Pipe 3- STD
Pipe64- STD	8.62	1,040	73.2	62.7	4.80	0.604	22.5	16.6	24.2	Pipe 2–1/2- STD
Pipe51- STD	5.44	658	60.5	52.5	3.63	0.261	11.7	8.65	20.1	Pipe 2- STD
Pipe38- STD	4.04	483	48.3	40.9	3.43	0.122	6.90	5.06	15.9	Pipe 1–1/2- STD
Pipe32- STD	3.38	403	42.2	35.1	3.30	0.0766	5.00	3.64	13.8	Pipe 1–1/4- STD

Table A1.5 *continued*

Section	*W*	*A*	*OD*	*ID*	t_{des}	*I / 10^6^*	*Z / 10^3^*	*S / 10^3^*	r	*Section*
(Metric)	*(kg/m)*	*(mm^2)*	*(mm)*	*(mm)*	*(mm)*	*(mm^4)*	*(mm^3)*	*(mm^3)*	*(mm)*	*(Imperial)*
Extra Strong										
Pipe310-XStrong	97.4	11,300	325	300	11.8	141	1,150	872	110	Pipe12-XStrong
Pipe254-XStrong	81.5	9,740	274	248	11.8	82.8	806	606	92.5	Pipe10-XStrong
Pipe203-XStrong	64.5	7,680	219	194	11.8	41.6	508	379	73.4	Pipe8-XStrong
Pipe152-XStrong	42.5	5,050	168	146	10.2	15.9	256	190	55.9	Pipe6-XStrong
Pipe127-XStrong	30.9	3,700	141	122	8.86	8.12	156	115	47.0	Pipe5-XStrong
Pipe102-XStrong	22.3	2,670	114	97.3	8.00	3.80	90.6	66.4	37.6	Pipe4-XStrong
Pipe75-XStrong	15.3	1,830	88.9	73.7	7.11	1.54	47.7	34.6	29.0	Pipe3-XStrong

Source: AISC Steel Construction Manual, 14th edition

1) Double extra strong sections not shown. See Steel manual.
2) $I/10^6$, multiply table value by 10^6
3) Z or $S/10^3$, multiply table value by 10^3

Beam Solutions

Appendix 2

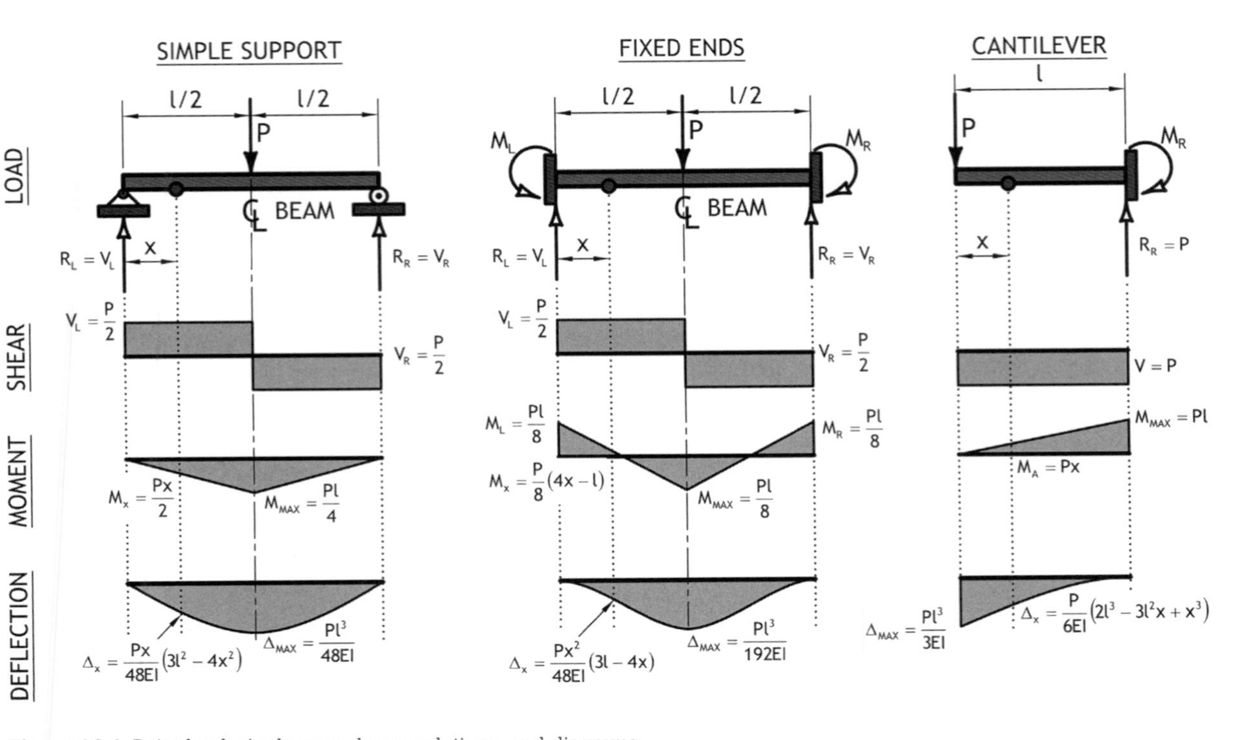

Figure A2.1 Point load, single-span, beam solutions, and diagrams

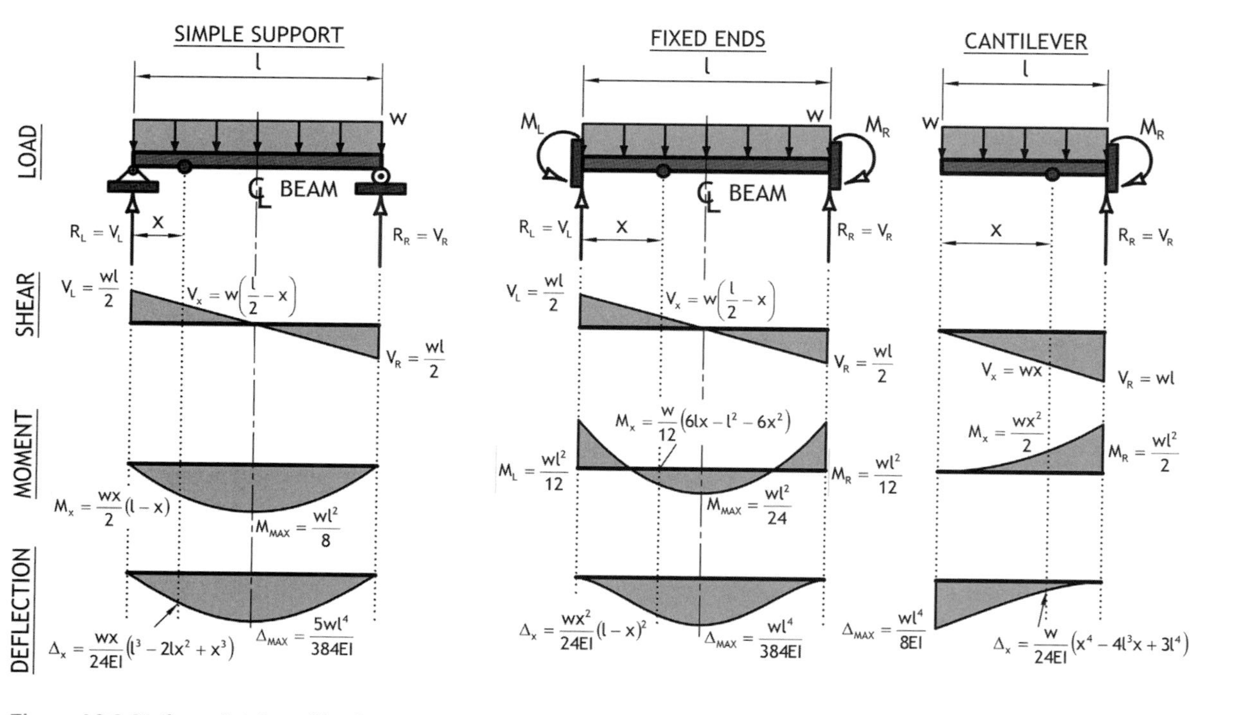

Figure A2.2 Uniform distributed load, single span, beam solutions, and diagrams

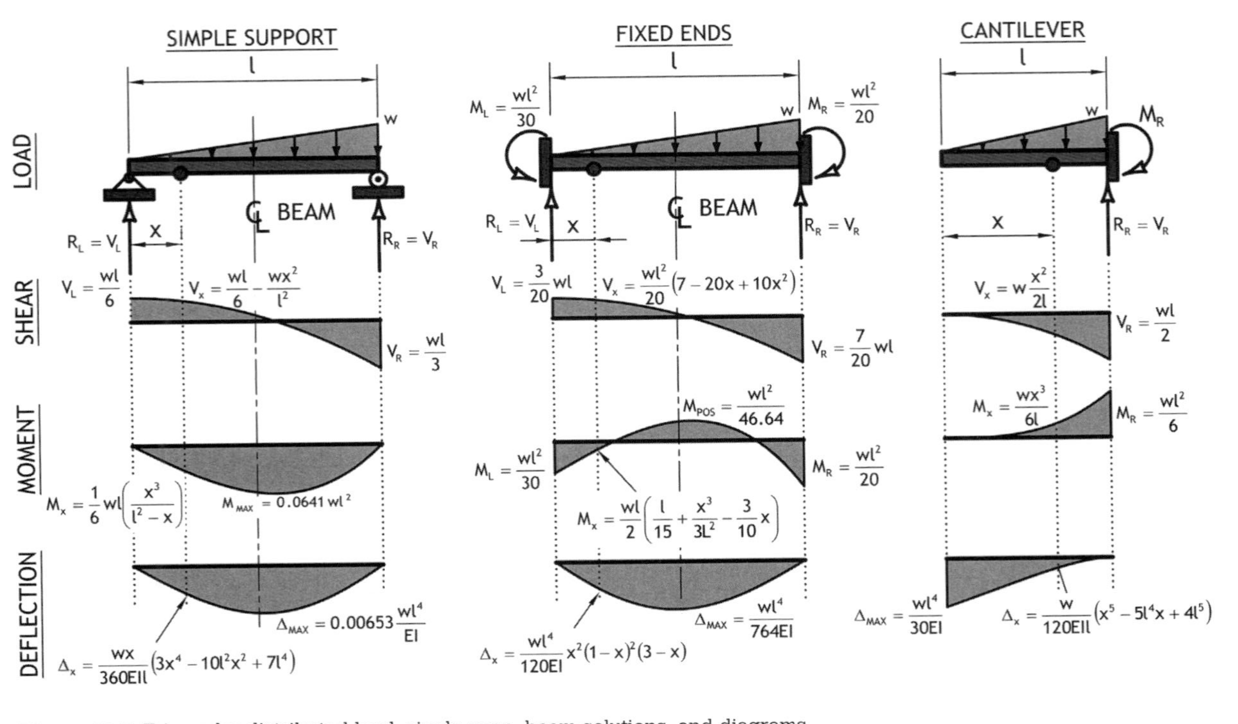

Figure A2.3 Triangular distributed load, single-span, beam solutions, and diagrams

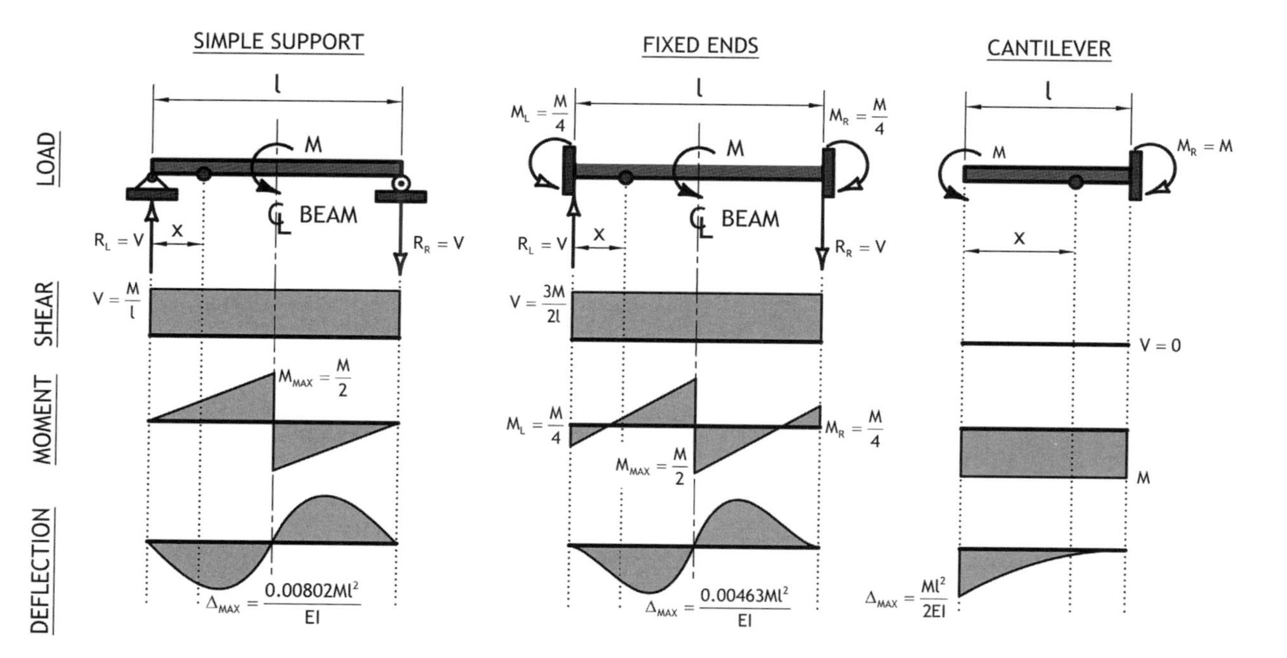

Figure A2.4 Moment load, single span, beam solutions, and diagrams

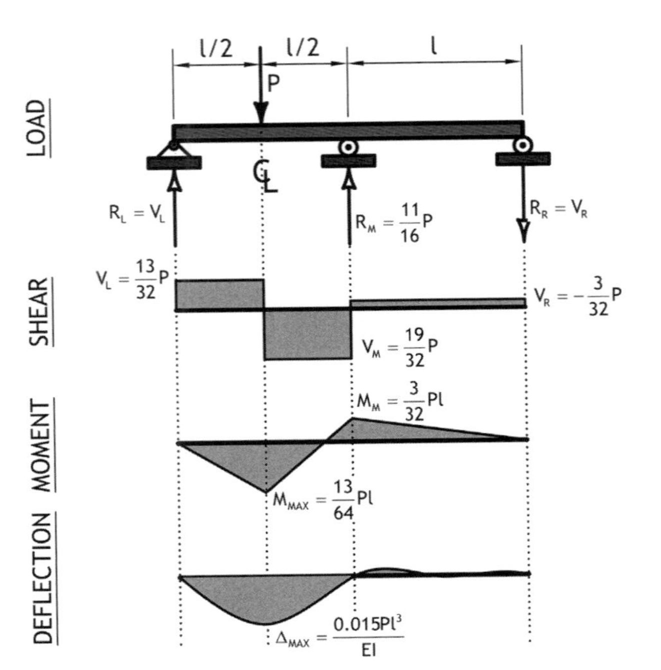

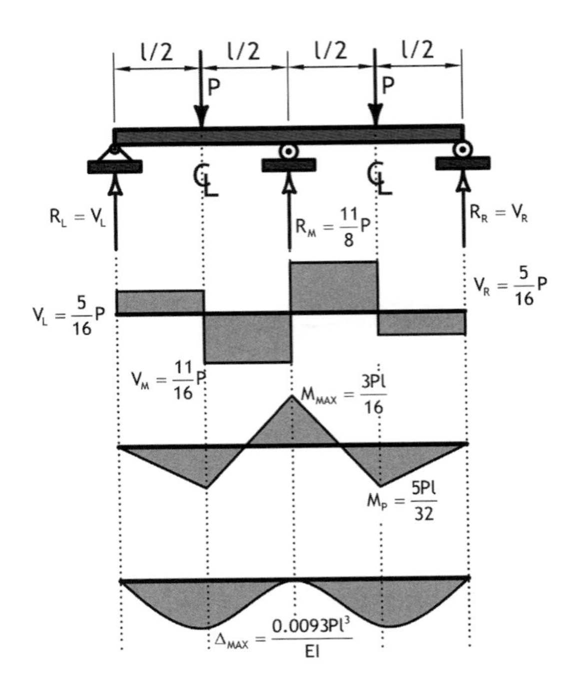

Figure A2.5 Point load, double-span, beam solutions, and diagrams

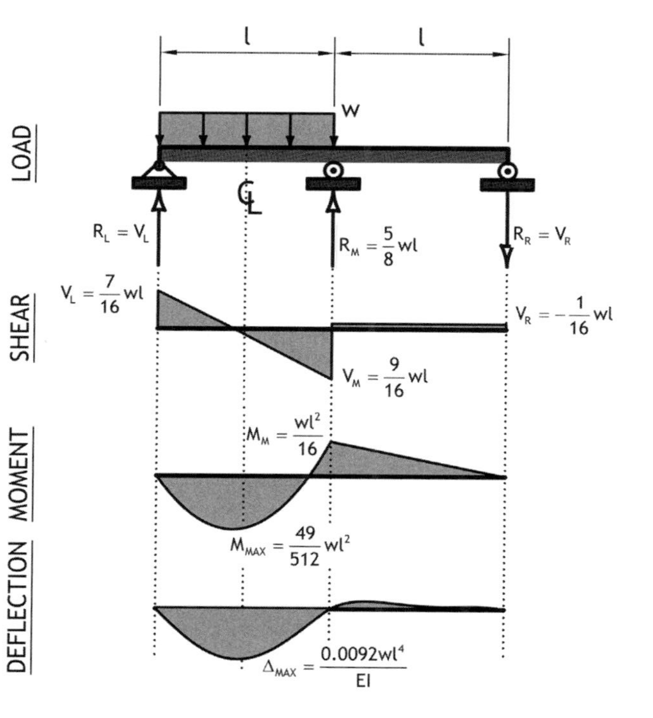

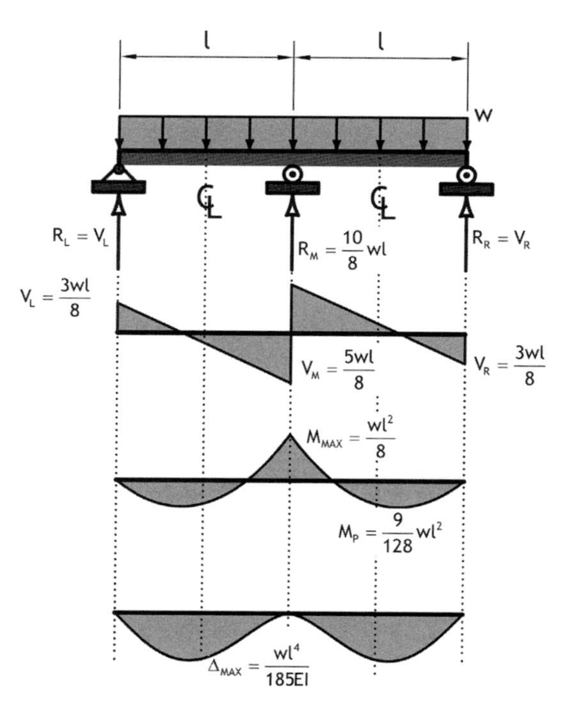

Figure A2.6 Uniform distributed load, double-span, beam solutions, and diagrams

Units

Appendix 3

Table A7.1 Units

Imperial Units		
Units	*Definition*	*Typical Use*
°	degrees	angle
deg	degrees	angle
ft	feet	length
ft^2	square feet	area
ft^3	cubic feet	volume
hr	hour	time
in	inches	length
in^2	square inch	area
in^3	cubic inch	volume
in^4	inches to the fourth power	moment of inertia
k	kip (1000 pounds)	force
k/ft	kips per foot (aka klf)	distributed linear force
k/ft^2	kips per square foot (aka ksf)	distributed area force, pressure
k/ft^3	kips per cubic foot (aka kcf)	density
k/in^2	kips per square inch (aka ksi)	distributed area force, pressure
k-ft	kip-feet	moment, torque
lb, lb_f	pound	force
lb/ft	pounds per foot (aka plf)	distributed linear force
lb/ft^2	pounds per square foot (aka psf)	distributed area force, pressure
lb/ft^3	pounds per cubic foot (aka pcf)	density
lb/in^2	pounds per square inch (aka psi)	distributed area force, pressure
lb-ft	pound-feet	moment, torque
rad	radian	angle
yd^3	cubic yard	volume

Metric Units		
Unit	*Definition*	*Typical Use*
°	degrees	angle
deg	degrees	angle
g	gram	mass
hr	hour	time
kN	newton	force
kN	kiloNewton	force
kN/m	kiloNewton per meter	distributed linear force
kN/m^2	kiloNewton per square meter (aka kPa)	distributed area force, pressure
kN/m^3	kiloNewton per cubic foot	density
kN-m	kiloNewton-meter	moment, torque
m	meters	length
m^2	square meters	area
m^3	cubic meters	volume
min	minute	time
mm	millimeters	length
mm^2	square millimeters	area
mm^3	cubic millimeters	volume
mm^4	millimeters to the fourth power	moment of inertia
MN/m^2	kiloNewton per square millimeter (aka MPa)	distributed area force, pressure
N	newton	force
N/m	newtons per meter	distributed linear force
N/m^2	newtons per square meter (aka MPa)	distributed area force, pressure
N/m^3	newtons per cubic meter	density
N/mm^2	newtons per square millimeter (aka MPa)	distributed area force, pressure
Pa	newton per square meter (N/m^2)	distributed area force, pressure
rad	radian	angle

Symbols

Appendix 4

Table A8.1 Symbols

Symbol	*Definition*	**Imperial**	**Metric**
$\#_b$	bending (flexure) property or action	vary	
$\#_c$	compression property or action	vary	
$\#_D$	dead load-related action	vary	
$\#_L$	live load-related action	vary	
$\#_n$	nominal capacity	vary	
$\#_S$	snow load-related action	vary	
$\#_t$	tension property or action	vary	
$\#_u$	factored load, any type	vary	
$\#_v$	shear property or action	vary	
$\#_W$	wind load-related action	vary	
a	fillet weld leg size	in	mm
A	area	in^2	mm^2
A_b	nominal unthreaded bolt cross-section	area	in^2
mm^2	A_e	effective area	in^2
mm^2	A_g	gross cross	section area
in^2	mm^2	A_{gv}	gross shear
area	in^2	mm^2	A_n
net cross	section area	in^2	mm^2
A_{nt}	net tension area	in^2	mm^2
A_{nv}	net shear area	in^2	mm^2
A_{req}	required area	in^2	mm^2
A_w	web area	in^2	mm^2
A_{we}	effective weld area	in^2	mm^2
B	width of HSS	in	mm
b	width of unstiffened compression element	in	mm
b'	material cantilever in prying	in	mm
b_f	flange width	in	mm
C	compression	lb or k	N, kN, MN

Symbol	*Definition*	**Imperial**	**Metric**
C	HSS torsion parameter	in^3	mm^3
C_b	lateral torsional buckling factor	unitless	
CE	carbon equivalent	%	
C_v	web shear buckling coefficient	unitless	
C_w	torsion warping factor	unitless	
d	nominal bolt diameter	in	mm
d	depth	in	mm
D	dead load	k, k/ft, k/ft^2 lb, lb/ft, lb/ft^2	kN, kN/m, kN/m^2 N, N/m, N/m^2
E	modulus of elasticity	lb/in^2	GN/m^2, GPa
E	seismic load	lb or k	N, kN, MN
e	eccentricity	in	mm
F	stress	lb/in^2	MN/m^2, MPa
F_{cr}	critical stress	lb/in^2	MN/m^2, MPa
F_e	elastic buckling stress	lb/in^2	MN/m^2, MPa
F_{EXX}	weld filler metal strength classification	lb/in^2	MN/m^2, MPa
F_n	nominal fastener stress	lb/in^2	MN/m^2, MPa
F_{nt}	nominal fastener tensile stress	lb/in^2	MN/m^2, MPa
F_{nv}	nominal fastener shear stress	lb/in^2	MN/m^2, MPa
F_{nw}	nominal weld metal stress	lb/in^2	MN/m^2, MPa
F_u	nominal tensile stress	lb/in^2	MN/m^2, MPa
F_y	nominal yield stress	lb/in^2	MN/m^2, MPa
g	gage	in	mm
G	shear modulus of elasticity	lb/in^2	MN/m^2, MPa
h	story height	ft	m
h	clear distance between flanges minus fillets	in	mm
H	HSS depth	in	mm
h_o	distance between flange centroids	in	mm

Symbol	*Definition*	**Imperial**	**Metric**
I, I_x, I_y	moment of inertia	in^4	mm^4
ID	inside diameter	in	mm
J	torsion factor	in^4	mm^4
K	effective length factor	unitless	
k_v	shear buckling coefficient	unitless	
L	live load	k, k/ft, k/ft^2 lb, lb/ft, lb/ft^2	kN, kN/m, kN/m^2, N, N/m N/m^2
L	span length	ft	m
l	connection length	in	mm
L_b	length between points of lateral beam bracing	ft	m
L_c	effective length factor	in	mm
l_c	clear distance between edge of material and hole	in	mm
L_o	base live load	k, k/ft, k/ft^2, lb lb/ft, lb/ft^2	kN, kN/m, kN/m^2, N N/m, N/m^2
L_p	limiting laterally unbraced length for plastic flexural yielding	ft	m
L_r	limiting laterally unbraced length for elastic flexural yielding	ft	m
L_t	tributary width	ft	m
l_w	weld length	in	mm
M_p	plastic flexural capacity	k-ft	kN-m
M_r	flextural capacity at elastic transition	k-ft	kN-m
M_u	flexural moment demand	k-ft	kN-m
M_y	moment at yield	k-ft	kN-m
n	number, quantity	unitless	
OD	outside diameter	in	mm
P	point load, axial compression	k	kN, MN
p	tributary length to fastener	in	mm
P_n	nominal compression capacity	k	kN, MN

Symbol	*Definition*	**Imperial**	**Metric**
P_u	factored compression demand	k	kN, MN
q, q_x	area unit load, pressure	lb/ft^2, k/ft^2	N/m^2, kN/m^2
r	radius of a circle or cylinder	in, ft	mm, m
R	response modification factor for seismic	force	unitless
R, $R_{\#}$	reaction	lb or k	N, kN, MN
R_n	nominal connection strength	k	kN, MN
R_t	ratio of expected to specified ultimate	strength	unitless
r_{ts}	effective radius of gyration	in	mm
r_x, r_y, r_z	radius of gyration	in	mm
R_y	ratio of expected to specified yield strength	unitless	
S	snow load	k, k/ft, k/ft^2, lb, lb/ft, lb/ft^2	kN, kN/m, kN/m^2, N, N/m, N/m^2
s	spacing	in	mm
S, S_x, S_y	elastic section modulus	in^3	mm^3
T	tension	lb or k	N, kN, MN
t	thickness	in	mm
t_{des}	design thickness	in	mm
t_e	fillet weld throat	in	mm
t_f	flange thickness	in	mm
t_{min}	minimum thickness to ignore prying	in	mm
T_n	nominal tension capacity	k	kN
T_u	factored tension demand	k	kN
T_u	factored torsion demand	k-ft	kN-m
t_w	web thickness	in	mm
U	shear lag factor	unitless	
U_{bs}	shear lag factor for block shear	unitless	
V	shear	lb or k	N, kN, MN
V_n	nominal shear capacity	k	kN
V_u	factored shear demand	k	kN

Symbol	*Definition*	**Imperial**	**Metric**
v_u	unit shear	lb/ft	kN/m
w	width	in	mm
w	line load, or uniform load	lb/ft	kN/m
W	wind load	k, k/ft, k/ft^2, lb, lb/ft, lb/ft^2	kN, kN/m, kN/m^2, N, N/m, N/m^2
W	weight	lb or k	N, kN, MN
W	diaphragm width	ft	m
w_D	line dead load	lb/ft	kN/m
w_L	line live load	lb/ft	kN/m
w_S	line snow load	lb/ft	kN/m
w_u	factored line load	lb/ft	kN/m
x	geometric axis, distance along axis	unitless	
$\bar{x}$	connection eccentricity	in	mm
y	geometric axis, distance along axis	unitless	
z	geometric axis, distance along axis	unitless	
Z, Z_x, Z_y	plastic section modulus	in^3	m^3
Δ	drift	in	mm
δ	deflection	in	mm
δ_a	allowable deflection	in	mm
ε	strain	unitless	
ϕ	strength reduction factor	unitless	
γ	unit weight	lb/ft^3	kN/m^3
λ_r	plastic slenderness parameter	unitless	
λ_p	compact slenderness parameter	unitless	

1) # indicates a general case of symbol and subscript, or subscript and symbol. It can be replaced with a letter or number, depending on how you want to use it. For example $R_{\#}$ may become R_V for a vertical reaction. Similarly, $\#_c$ may become P_c, indicating a compressive point load.

Units Conversions

Appendix 5

Table A5.1 Unit conversion table

Imperial to Metric					
Multiply	ft	by	0.305	to get	m
	ft^2		0.093		m^2
	ft^3		0.028		m^3
	in		25.4		mm
	in^2		645.2		mm^2
	in^3		16,387		mm^3
	in^4		416,231		mm^4
	k		4.448		kN
	k/ft		14.59		kN/m
	k/ft^2		47.88		kN/m^2
	k/ft^3		157.1		kN/m^3
	k/in^2 (ksi)		6.895		MN/m^2 (MPa)
	k-ft		1.356		kN-m
	lb, lb_f		4.448		N
	lb/ft		14.59		N/m
	lb/ft^2(psf)		47.88		N/m^2 (Pa)
	lb/ft^3		0.157		kN/m^3
	lb/in^2		6,894.8		N/m^2
	lb-ft		1.355		N-m
	lb_m		0.454		kg

SI to Imperial					
Multiply	m	by	3.279	to get	ft
	m^2		10.75		ft^2
	m^3		35.25		ft^3
	mm		0.039		in
	mm^2		0.0016		in^2
	mm^3		6.10237E-05		in^3
	mm^4		2.40251E-06		in^4
	kN		0.225		k
	kN/m		0.069		k/ft
	kN/m^2		0.021		k/ft^2
	kN/m^3		0.0064		k/ft^3
	MN/m^2 (MPa)		0.145		k/in^2 (ksi)
	kN-m		0.738		k-ft
	N		0.225		lb, lb_f
	N/m		0.069		lb/ft
	N/m^2 (Pa)		0.021		lb/ft^2(psf)
	kN/m^3		6.37		lb/ft^3
	N/m^2		1.45E-04		lb/in^2
	N-m		0.738		lb-ft
	kg		2.205		lb_m

Bibliography

AISC. *Code of Standard Practice for Steel Buildings and Bridges*, AISC 303 (Chicago: American Institute of Steel Construction, 2016).

AISC. *Seismic Provisions for Structural Steel Buildings*, AISC 314 (Chicago: Research Council on Structural Connections, 2016).

AISC. *Specification for Structural Joints using High-Strength Bolts* (Chicago: American Institute of Steel Construction, 2009).

AISC. *Specification for Structural Steel Buildings,* AISC 360 (Chicago: American Institute of Steel Construction, 2016).

AISC, *Steel Construction Manual*, 14th Edition (Chicago: American Institute of Steel Construction, 2011).

ASCE. 2010. *Minimum design loads for buildings and other structures*, ASCE/SEI 7–10 (Reston: American Society of Civil Engineers).

AWS. *Structural Welding Code—Steel*, D1.1 (Miami: American Welding Society, 2015).

AWS. *Structural Welding Code—Sheet Metal*, D1.3 (Miami: American Welding Society, 2008).

AWS. *Structural Welding Code—Reinforcing Steel*, D1.4 (Miami: American Welding Society, 2011).

AWS. *Bridge Welding Code*, D1.5 (Miami: American Welding Society, 2015).

AWS. *Structural Welding Seismic,* D1.8 (Miami: American Welding Society, 2016).

Blodgett, O. W. *Design of Welded Structures*, 12th printing (Ohio: Lincoln Arc Welding Foundation, 1982).

Cook, R.D., Young, W.C. *Advanced Mechanics of Materials,* (Upper Saddle River: Prentice Hall, 1985).

Design Guide 11: Floor Vibrations Due To Human Activity, AISC/CISC Steel Design Guide Series 11, American Institute of Steel Construction, 1997.

Esra Hasanbas Persellin. "Vibration." In *Special Structural Topics*, edited by Paul W. McMullin and Jonathan S. Price. (New York: Routledge, 2017).

"GeoHistory Resources," Greater Philadelphia GeoHistory Network, accessed May 1, 2017. *www.philageohistory.org/rdic-images/index2.cfm.*

Holy Bible. Authorized King James Version, Judges 16.21–30 (Oxford UP, 1998).

ICC-ES, *Vulcraft Steel Deck Panels*, ESR-1227 (Brea, CA; ICC Evaluation Service, 2016).

RSSC. *Specification for Structural Joints Using High-Strength Joints*, (Chicago: Research Council on Structural Connections, 2014).

SAE. *Mechanical and Material Requirements for Externally Threaded Fasteners*, J429 (Warrendale: SAE International, 2014).

Seaburg, P.A., Carter, C.J. *Torsional Analysis of Structural Steel Members*, Steel Design Guide 9, (Chicago: American Institute of Steel Construction, 1997).

Teran Mitchell. "Structural Materials." In *Introduction to Structures*, edited by Paul W. McMullin and Jonathan S. Price. (New York: Routledge, 2016).

William Komlos. "Quality and Inspection." In *Special Structural Topics*, edited by Paul W. McMullin and Jonathan S. Price. (New York: Routledge, 2017).

Glossary

AISC	American Institute of Steel Construction
angle	L-shaped section, common in bracing and connections
arc weld	welding process utilizing an electric arc to melt base and filler metals
area load	load applied over an area
area, cross-sectional	area of member when cut perpendicular to its longitudinal axis
ASD	allowable stress design; factors of safety are applied to the material strength
aspect ratio	ratio of two perpendicular dimensions
axial	action along length (long axis) of member
axis	straight line that a body rotates around, or about which a body is symmetrical
base shear	horizontal shear at base of structure due to lateral wind or seismic forces
bay	space between columns
beam	horizontal member resisting forces through bending
bearing	compressive forces transmitted by two members in direct contact
bearing wall	wall that carries gravity loads
block shear	connection failure mode characterized by shear and tension failure

brace	member resisting axial loads (typically diagonal), supports other members
braced frame	structural frame whose lateral resistance comes from diagonal braces
brazing	low temperature metal joining process
buckling	excess deformation or collapse at loads below the material strength
buckling restrained brace	brace in concrete and steel jacket, with same tension and compression capacity
camber	intentional curvature in beams and trusses to offset deflection
cantilever	structural member projecting from a rigid support on only one end
capacity	ability to carry load, related to strength of a member
carbon equivalence	measure of steel weldability
channel	C-shaped section, used as light beams
charpy test	notch toughness test for quality control
chord	truss or diaphragm element resisting tension or compression forces
code	compilation of rules governing the design of buildings and other structures
collector	see drag strut
column	vertical member that primarily carries axial compression load, supports floors and roofs
compact	section capable of full yield prior to local buckling
complete joint penetration	groove weld through the entire joint thickness
component	single structural member or element
compression	act of pushing together, shortening
connection	region that joins two or more members (elements)
construction documents	written and graphical documents prepared to describe the location, design, materials, and

	physical characteristics required to construct the project
cope	flange cutout to fit within receiving member
core	central building area, stacking vertically, with stairs, elevators, and mechanical shafts, and heavy structure
couple, or force couple	parallel and equal, but opposite forces, separated by a distance
creep	slow, permanent material deformation under sustained load
curtain wall	glass and metal panel façade system
dead load	weight of permanent materials
deflection δ	movement of a member under load or settlement of a support
demand	internal force due to applied loads
depth	height of bending member, or larger dimension of column
design thickness	assumed wall thickness to account for manufacturing variations
detailing	process of preparing detailed piece drawings from engineering drawings
diaphragm	floor or roof slab transmitting forces in its plane to vertical lateral elements
discontinuity	interruption in material
distributed load	line load applied along the length of a member
doubler	plate added to a flange or web to increase strength
drag strut	element that collects diaphragm shear and delivers it to a vertical lateral element
drift	lateral displacement between adjacent floor levels in a structure
durability	ability to resist deterioration
eccentric braced frame	braced frame with workpoints offset, causing yielding in beam to dissipate energy

eccentricity	offset of force from centerline of a member, or centroid of a fastener group
edge distance	distance between fastener center line and edge of member
effective area	net area adjusted to account for shear lag
effective length	length of member in compression, accounting for end restraint
effective length factor	factor to adjust member length, based on end restraint conditions
elastic	ability to return to original shape after being loaded
element	single structural member or part
empirical design	design methodology based on rules of thumb or past experience
end distance	distance between fastener centerline and end of member
erection drawings	drawings showing each piece of steel, it's mark, and location in the structure
expansion joint	separation between adjacent parts of the structure, to allow relative movement and avoid cracking
expected strength	actual material strength, usually above specified strength
fatigue	time-dependent crack growth due to cyclic stresses
faying surfaces	surfaces in contact in a connection
field weld	weld that may or is required to be made in the field
filler metal	weld metal used to fill gaps or build up welds
fillet weld	weld made at the intersection between two elements
fixed	boundary condition that does not permit translation or rotation
flare bevel weld	weld made at the intersection between a planar and curved surface

flexibility factors	stiffness measurement in metal deck diaphragms
flexure, flexural	another word for bending behavior
footing	foundation system bearing on soil near the ground surface
force	effect exerted on a body
forge welding	welding process by folding and hammering material causing it to overlap and mix
frame	system of beams, columns, and braces, designed to resist vertical and lateral loads
free body diagram	elementary sketch showing forces acting on a body
gage	center-to-center spacing of fasteners
girder	beam that supports other beams
gouge	removal of material by mechanical means or high temperature
gravity load	weight of an object or structure, directed to the center of the earth
grip	thickness of material that a bolt joins
groove weld	weld placed in a groove between connected elements
gusset plate	plate used to connect members, typically in trusses or braced frames
high-strength bolt	bolt with tensile strength above 100 k/in2 (690 MN/m2)
hollow structural section	round or rectangular section, hollow in the middle, common as braces and columns
HSS	hollow structural section, rectangular or round member
indeterminate	problem that cannot be solved using the rules of static equilibrium alone, number of unknowns greater than number of static equilibrium equations
inelastic	behavior that goes past yield, resulting in permanent deformation

irregularity	abrupt change in structure geometry causing force concentrations
joint	area where two or more members ends are connected
lap joint	joint made by overlapping two connection elements
lateral bracing	bracing intended to prevent buckling or lateral torsional buckling
lateral load	load applied in the horizontal direction, perpendicular to the pull of gravity
lateral torsional buckling	condition where beam rolls over near the middle due to inadequate bracing for the given load
live load	load from occupants or moveable building contents
live load reduction	code permitted reduction when area supported by a single element is sufficiently large
load	force applied to a structure
load combination	expression combining loads that act together
load factors	factor applied to loads to account for load uncertainty
load path	route a load takes through a structure to reach the ground
local buckling	buckling of a portion of a member cross section
LRFD	load and resistance factor design, also called strength design
modulus of elasticity E	material stiffness parameter, measure of a material's tendency to deform when stressed
moment arm	distance a force acts from a support point
moment connection	connection that transmits bending moments between members
moment frame	structural frame whose lateral resistance comes from rigid beam-column joints

moment M	twisting force, product of force and the distance to a point of rotation
moment of inertia I	geometric bending stiffness parameter, property relating area and its distance from the neutral axis
net area	area accounting for removal of material due to holes or slots
neutral axis	axis at which there is no lengthwise stress or strain, point of maximum shear stress or strain, neutral axis does not change length under load
nominal dimensions	theoretical dimension, often used in tables to identify members
nominal strength	element strength, typically at ultimate level, prior to safety factor application
nondestructive testing	inspection procedure that leaves the part undamaged, like ultrasonic testing
partial joint penetration	groove weld through part of the joint thickness
P-delta	increased moment due to axial load and deflection
pin	boundary condition that allows rotation but not translation
pitch	lengthwise spacing of fasteners
plane strain	condition where strain is zero in one direction, perpendicular to applied stress, associated with thicker material
plastic	occurs after yield, where material experiences permanent deformation after load is removed
plastic section modulus	section property accounting for full section yielding
plug weld	weld made in a hole or slot
point load	concentrated load applied at a discrete location
point of inflection	point in deflected shape where there is no moment, deflected shape changes direction

pressure	force per unit area
pretensioned bolt	bolt tightened to a high percentage of its tensile strength
protected zone	member area where no connections are permitted
prying	increased tensile force in a bolt due to plate bending
radius of gyration	relationship between area and moment of inertia used to predict buckling strength
reaction	force resisting applied loads at end of member or bottom of structure
rivets	fastener created by hammering a hot, round bar to form a head
roller	boundary condition that allows rotation, but limits translation in only one direction
row of fasteners	fasteners in a line parallel to the applied force
rupture	complete separation of material
safety factor	factor taking into account material strength or load variability
section modulus	geometric bending strength parameter
seismic design category	classification based on occupancy and earthquake severity
seismic load	force accounting for the dynamic response of a structure or element due to an earthquake
seismic response modification factor	factor reducing seismic force based on energy dissipation ability of the structure
seismic-force resisting system	portion of structure designed to resist earthquake effects
service load	unfactored loads, used for checking deflection
shear	equal, but opposite forces or stresses, acting to separate or cleave a material, like scissors
shear lag	internal shear stress when connection is only on part of a cross-section, creates non-uniform tensile stress in a member

shear plate	high-strength connection made from placing a steel plate into a round groove, thereby engaging a large surface area
shear wall	wall providing lateral resistance for structure
shop drawings	drawings of each piece of steel, showing all information required to fabricate it
sign convention	method of assigning positive and negative values to the direction of loads, reactions, and moments
simple connection	connection that transmits negligible moment
simplifying assumption	assumption that makes the problem easier to solve, but is realistic
slender	member or element that is prone to buckling
slip critical connection	connection utilizing pretensioned bolts, transferring force through friction
snow load	load from fallen or drifted snow
snug tightened bolt	bolt tightened with full effort of ironworker on a spud wrench, keeping faying surfaces in contact
soft story	story with substantially lower stiffness than adjacent stories
spacing	center-to-center distance between adjacent items
span length	clear distance between supports
special seismic systems	structural systems specifically detailed to absorb seismic energy through yielding
splice	connection between ends of the same element
stability	structure's resistance to excessive deformation or collapse at loads below the material strength, opposite of buckling
stiffener	plate welded into structural member to increase local strength
stiffness	resistance to deformation when loaded

strain	change in length divided by initial length, percent change in length if multiplied by 100
strength	material or element resistance to load or stress
strength design	load and resistance factor design; safety factors applied to the loads and materials
strength reduction factor	LRFD material safety factor
structural analysis	determination of forces, moments, shears, torsion, reactions, and deformations due to applied loads
structural integrity	ability of structure to redistribute forces to maintain overall stability after localized damage occurs
structural system	series of structural elements (beams, columns, slabs, walls, footings) working together to resist loads
support	either the earth or another element that resists movement of the loaded structure or element
sway frame	flexible frame
tension	act of pulling apart, lengthening
tension field action	diagonal tension region due to shear forces, and compression buckling
torsion	act of twisting along an axis
toughness	material resistance to cracking
tributary area	area supported by a structural member
tributary width	width supported by a structural member, usually a beam, joist or girt
truss	structural member comprised of axially loaded members arranged in triangular fashion
unbraced length	length between brace point where a member can buckle
weak axis	axis with lower strength properties, typically the y-axis

weathering steel	steel that forms a tight oxide layer, substantially slowing corrosion in non-marine environments
welding	fusion of metals by melting them together, often with a filler metal
wide flange	I-shaped section, with wide flanges, commonly used in beams and columns
width	smaller member dimension
wind load	force due to wind
yield	point at which a material has permanent deformation due to applied loads, start of inelastic region of stress-strain curve

Index